*Secrets of the Spas*

# Secrets of the Spas

PAMPER AND VITALIZE YOURSELF AT HOME

*By Catherine Bardey*

Photographs by Zeva Oelbaum

BLACK DOG
& LEVENTHAL
PUBLISHERS

Published by
Black Dog & Leventhal Publishers, Inc.
151 West 19th Street
New York, NY  10011

Distributed by Workman Publishing Company
708 Broadway
New York, NY 10003

Printed in Great Britain

ISBN 1-57912-063-6
h g f e d c b a
Health and body images provided by Digital Vision™
Library of Congress Cataloging-in-Publication Data

Bardey, Catherine, 1963-
Secrets of the Spas: pamper and vitalize yourself at home:
lotions, potions, oils, rubs, scrubs, wraps, masks & cuisine
by Catherine Bardey.
p. cm.
Includes bibliography.

ISBN 1-57912-063-6

1. Soap.  2. Cosmetics.  3. Beauty.  4. Personal.  I. Title.

TP983.B293    1999
668' .55—dc21     98-50121

Design by 27.12 Design, Ltd.
Studio photographs by Zeva Oelbaum

# Table of Contents

INTRODUCTION 7

CHAPTER 1 *Body* 9

    Body Scrubs 10

    Body Wraps 19

    Herbal Baths 27

    Essential Oil Baths 32

    Moisturizers 47

    Massage 55

CHAPTER 2 *Face* 63

    Cleansers 67

    Steams 73

    Masks and Peels 78

    Toners 94

    Moisturizers 101

CHAPTER 3 *Hands & Feet* 107

    Cuticles and Nails 108

    Hand and Foot Scrubs, Exfoliators & Washes 114

    Moisturizers 127

Chapter 4 *Hair* 129

    Pre-Shampoo Treatments 130

    Shampoos and Conditioners 139

    Rinses 142

Chapter 5 *Spa Cuisine* 147

    Drinks 149

    Soup and Salads 152

    Desserts 156

Chapter 6 *Essentials* 159

    Oils—Essential and Carrier 160

    Pampering Tools 168

    Glossary 172

    Associations, Resources and Suppliers 179

    Spa Listings 182

    Shopping Guide 187

    Bibliography 188

    Conversion Table 190

INDEX 191

# Secrets of the Spas

**V**ery rarely do we take time out of our busy schedules to sit back and indulge ourselves for a few minutes. Usually, a soothing cucumber mask or an oatmeal, almond and avocado body scrub are fleeting fantasies during a stressful day. As soon as a stoplight turns green or a telephone shrills then we must snap out of the calming thoughts of spa indulgences and return to our daily grind.

*Secrets of the Spas* is the ultimate treasure full of recipes and hints on how to revitalize and indulge yourselves—both body and mind—at home.

*Secrets of the Spas*, unlike many other spa books, caters to every woman. Whether you have the time to go to the store and buy treatments, make them yourself, or take the weekend off and get away—this book is for you.

The woman who can't get away to a spa for a weekend or even a few hours can luxuriate at home with recipes for lotions, oils, rubs, scrubs, wraps, masks and cuisine. The book gives in-depth instruction not only on how to whip up the recipes—but do so in no time, for pennies, while recreating a spa setting in your own home.

If you you're not inclined to personally make the recipes, familiarize yourself with the various treatments in the book and go to the suggested department stores, cosmetic, body or specialty shops nearest you. Purchase whichever products suit your body and skin type and use them in the privacy of your own home, whenever your busy schedule permits.

If you have had great success in creating your home spa and want to experience it from another perspective, avail yourself of the international spa listing and go elsewhere to be pampered. This chart contains hundreds of spas from all over the world, each individually coded with its own spa specialty. The spas listed fall into one or more of the following categories: wellness, adventure, new age/holistic, culinary, water treatment, fitness and women only.

No matter what type of indulgence you are looking for, *Secrets of the Spas*, holds your hand at every level. All you need to do is get comfortable and dive in—whether you are in a bookstore, on a train or in your office—you are never too far from being swept away into pure, luxurious, spa-induced satisfaction (even if it is for just 30 minutes of your day).

# CHAPTER I

# Body

BODY SCRUBS

•

BODY WRAPS

•

BATHS

•

MOISTURIZERS

•

MASSAGE OILS & RUBS

•

# Body Scrubs

Imagine being bathed in purifying, scented waters, then snuggly wrapped in soft, warm linens. Imagine your mind drifting away to serenity while your body inhales the perfumes of the calming and detoxifying splendors, discovering a renewed self, revitalized, regenerated and fully soothed.

Exfoliation is an integral part of the spa treatment. Techniques might vary but the purpose of exfoliation is twofold: first, it rids the skin's surface of any dirt or oil residue and rubs off the uppermost layer of dead skin cells. Second, it prepares the skin for subsequent treatments. Removal of the layer of dead cells enables the skin to more effectively absorb or react to products utilized in treatments that follow the exfoliation process.

The treatments in this chapter, perhaps a bit more time-consuming than some of the facial or hair treatments, are well worth the indulgence. In addition to the physical benefits brought forth by the different materials and ingredients used, the effects on the spirit are equally rewarding. So go ahead, splurge into some seaweed and emerge with a new set of wings!

# Body-scrubbing is a multi-purpose treatment:

- It cleanses and exfoliates the outer layers of the skin by removing dirt and dead cells.

- It helps stimulate and circulate lymph fluids which aid in the elimination of toxins.

- It serves as a stimulating massage and improves blood circulation.

- It affects the secretion and production of the body's natural oils.

- The entire body should be scrubbed on a daily basis, even if only for 5 minutes.

- First-time scrubbers should be gentle on the skin at the beginning; do not abrade or irritate it. Gradually the skin will develop "resistance" to the scrubbing and more vigorous pressure can be applied.

- Make sure you have the appropriate body-scrubbing tools (remember—it's body scrubbing, not face scrubbing!): sisal mitt, towel, loofah, abrasive sponge or stiff brush *(see "Pampering Tools" page 168).*

- The scrubber should be used in a circular motion, starting with the feet, ankles and legs, then working on to the hands, arms and shoulders, gradually working your way down the back, then up through the torso towards the heart.

- Make sure you thoroughly wash your scrubbing tools with soap and water and hang them up to dry after each use—they can develop bacteria very quickly if not properly cleaned and dried after each use.

- A cool-water rinse after a body scrub will further activate blood circulation throughout the body. Pat dry and moisturize.

- As this treatment is revitalizing and energizing, it is best practiced in the morning.

# Revitalizing Sea Salt & Rosemary Oil Body Scrub

Ideal for dry and flaky skin, this scrub is both easy and fun to apply, especially if you do it with a partner and focus on each other's back. Sea salt exfoliates the skin's surface and rids it of dead cells; the tonic and invigorating effects of rosemary will increase blood flow.

## Recipe:

___

1/2 cup coarse sea salt
1/4 cup rosemary oil
body-scrub tool

*Dip the body-scrub tool in a bit of oil, then dip it in the coarse salt. Rub over skin in a circular motion, paying attention to rough spots like the knees, elbows and heels. Rinse off with warm water, pat dry and moisturize.*

•

*Rosemary revives, clears the mind and is commonly used in inhalations for nasal congestion.*

# Oatmeal, Almond & Avocado Body Scrub

Oatmeal, rich in vitamin E, is well-known for its anti-inflammatory, anti-itch, skin-soothing properties. Almonds are a natural emollient and, when ground, an effective exfoliant. Combined with avocado pulp, which contains vitamins A, D and E, and which possesses collagen-rebuilding properties, the oatmeal-almond mixture is ideal for irritated, tired and dry skin. In order to avoid clogging your drain, it is best to make a small pouch or bag out of muslin cloth and fill it with dry oatmeal and the grated almonds. Ideally, you should make the bag large enough so that you can fit your hand into it and use it as a mitt to rub the avocado on your body.

## Recipe:

—

1/2 cup almonds
1 cup dry oatmeal
1 ripe avocado, peeled, pitted and mashed

*Place almonds in a blender or coffee grinder and grind until coarse and not too fine.
Mix almonds and oatmeal and place in glove or muslin mitt. Scoop up avocado and rub on body with mitt. Rinse off with warm water, pat dry and moisturize.*

*Save the avocado pit as a natural massage tool and rub the pit into sore muscles after the scrub.*

# Exfoliating Sea Salt & Apricot Kernel Oil Body Scrub

This scrub will remove dead cells from the skin's surface and leave it remarkably soft and conditioned. Here, the salt serves a dual purpose in that its texture provides the slight abrasion or graininess necessary to "polish" the skin, increase blood circulation and stimulate lymph production to eliminate toxins, while naturally triggering the skin's own oil secretion. Apricot kernel oil, rich in natural emollients and vitamins A and B, is easily absorbed by the pores. It is also helpful in healing damaged skin cells and will leave your skin moisturized and smooth.

## Recipe:

—

2 cups coarse sea salt
1/4 cup apricot kernel oil
body-scrub tool

*Make a paste out of the sea salt and apricot kernel oil in a bowl. While standing in the bathtub, shower, or on a towel with a space heater nearby, place a handful of the mixture on your body (starting with the legs, then the arms, the torso, and the back) and begin spreading in a smooth, circular motion.*
*It is important to keep the palms of the hands and t he fingers flat against the surface of the skin while applying firm but not abrasive pressure to the skin.*
*The point of the rub is to affect the surface of the skin, not the underlying tissues. After the entire body has been scrubbed and exfoliated, gently shower or rinse off the mixture using a sponge or loofah and a mild bath wash (see page 27). Pat dry.*

# Make sure to replenish the body's fluids after the treatment by drinking plenty of water or herbal tea.

- Before the treatment, remove all lotions, perfumes and dirt from the skin with a quick shower.

- Heat retention is important, so remember to warm tools, towels and sheets as well as the wrap ingredients before the treatment.

- The ideal temperature of the water in which the cotton strips or sheet will be immersed is approximately 180 degrees F (82 degrees C).

- Wear rubber gloves when you wring out the sheet or strips so as to avoid burning your fingers.

- Place a space heater near the relaxation area so as to prolong the heat of the wrap for as long as possible.

- You can intensify the effects of the treatment and raise your body's temperature by sipping hot herbal tea before the wrap.

- After the wrap, put on a warm bathrobe and relax for 20 minutes. Do not attempt to do any physical activity immediately after the treatment.

*"The best and most beautiful things in the world cannot be seen or even touched. They must be felt."*

—ANONYMOUS

## Chamomile, Lavender & Valerian Relaxation Wrap

Valerian was traditionally used in ancient Rome as a mild sedative. Today, it is used as a mild sleep aid and anxiety reliever. Coupled with the calming and anti-spasmodic effects of chamomile and the restorative properties of lavender, this herbal wrap is pure bliss after a stressful day, especially before going to bed. It's best not to shower immediately after the wrap (and possibly not until the next morning) so as to prolong the effects of the moist heat.

•

*Lavender relaxes, soothes, restores and balances the system. It is excellent for aching feet, tired muscles and headaches.*

## Recipe:

—

3 cups fresh herbs:
1 cup chamomile, chopped
1 cup valerian root, grated
1 cup lavender, crushed or
1 1/2 cups dried herbs:
1/2 cup dried chamomile
1/2 cup dried valerian root
1/2cup dried lavender
1 cotton or linen sheet cut into 3–inch-wide strips
1 large pot for boiling *(lobster or spaghetti pot)*
1 plastic sheet or shower curtain liner

*First, select your relaxation spot (bed, lounge
chair or bathtub) and place the plastic sheet
or shower curtain on it. Set up the space heater
next to it and turn it on. Set blanket(s) nearby.
Place herbs in pot and fill with water. Bring
to a boil and simmer, covered, for 5 minutes.
Remove from heat and let sit for another 5
minutes. With a strainer, scoop out from the
water as many herbs as possible. Place sheet
strips in the pot and let steep for 5 minutes.
Remove strips from the pot one at a time as
you use them, and wring out excess water.
Start by snugly wrapping the strips around
legs, then arms, and finish with the torso.
Tuck the ends of the strip under a wrapped area
nearby so as to insure a snug fit. Lay yourself
on the plastic sheet or shower curtain, wrap
yourself in it, and cover your body with
blankets. Relax for 10 minutes, then slowly
remove strips. Roll into bed.*

# Remineralizing Seaweed Wrap

Seaweed wraps are exceptionally beneficial as a quick way to remineralize the body and replenish it with nutrients and elements lost on a daily basis—such as calcium, magnesium, potassium, iodine, protein, phosphorous, copper, vitamins and amino acids. The molecular formation of the human blood cell and that of seawater is similar, thereby enabling nutrients from the seawater to easily filter through the pores of the skin and directly into the bloodstream. The most popular type of seaweed used for cosmetic and healing purposes is kelp—the common name for the large, leafy brown algae that grows along colder coastlines. Although using fresh seaweed is certainly optimal, a seaweed powder mixed with enough water to create a mayonnaise-like paste will render equally satisfying results.

## Recipe:

—

8 ounces dried or powdered seaweed and
enough water to make a paste,
or approximately 3 ounces of fresh seaweed
Mylar or plastic sheet
1 blanket
Space heater (optional)

*Place a blanket at the bottom of the bathtub and set the Mylar or plastic sheet on top. Set the space heater nearby and turn it on. While sitting on the sheet, cover yourself with the seaweed paste or the fresh seaweed and then wrap yourself with the plastic sheet and blanket. Relax for 20 minutes. Slowly unwrap yourself and remove blanket and sheet from the tub. Fill the tub with warm water and relax for 10 minutes. After the treatment, pat skin dry and moisturize.*

# Ginger & Lemongrass Detoxification and Rejuvenation Wrap

Lemongrass' tonic, invigorating and antiseptic properties combined with ginger's stimulating and cleansing effects make this wrap an ideal morning ritual. Ginger, cultivated in Southeast Asia since 2,500 years ago for its medicinal properties, has also been found to rid the body of flu-like symptoms, colds and respiratory ailments. The citrus aroma of lemongrass will uplift and rejuvenate any mood.

## Recipe:

1 cup fresh lemongrass, chopped (or 1/2 cup dried)
1 3-inch-long piece of ginger root (grated or minced)
1 large pot for boiling (lobster or spaghetti pot)
1 cotton or linen sheet cut into 3-inch-wide strips
1 plastic sheet or shower curtain liner
1 or 2 blankets

*Place lemongrass and grated ginger root in a large pot and fill with water. Bring to a boil and simmer, covered, for 5 minutes. Remove from heat and let sit for another 5 minutes. With a strainer, scoop out as much ginger and lemongrass as possible. Place the sheet strips in the pot and let steep for 5 minutes. Remove strips from the pot one at a time as you use them, and wring out excess water. Start by snugly wrapping the strips around legs, then arms, and finish with the torso. Tuck the ends of the strip under a wrapped area nearby so as to insure a snug fit. Wrap yourself in the plastic sheet or shower curtain liner and relax in a lounge chair or bed. Cover yourself with blankets for warmth. After 15 minutes, remove sheet strips and shower.*

# The best time to bathe is when the stomach is empty.

- Salt added to bath water will help alleviate some of the light-headedness that some experience as a result of bathing in hot water.

- Morning baths should be tonic. Their primary purpose is to relieve physical fatigue, stimulate the body, and eliminate toxins. An ideal temperature for a morning bath is around 96.8 degrees F (36 degrees C).

- Evening baths are meant to be relaxing, to relieve stress and to prepare the body for sleep. An ideal temperature for an evening bath ranges between 98.6 degrees F (37 degrees C) and 102 degrees F (39 degrees C).

- A muscle- or joint-soothing bath after an intense physical exertion or a very stressful day generally ranges between 104 degrees F (40 degrees C) and 107 F (42 degrees C).

- Remember always to replenish body fluids by drinking plenty of water or herbal tea after a bath.

- After bathing, apply moisturizer while the skin is still warm and moist.

> *"The body is not a home but an inn, and that only briefly."*
>
> —SENECA

# Aromatic Herbal Baths

Herbal baths are detoxifying and aromatic and intensely relaxing, both spiritually and physically. The ceremony of bathing in herbs has become a daily rite in many cultures, a return to the source for the body and spirit.

The herb or combination of herbs used in the bath depends on the effect that is sought: to soothe the skin, promote sleep, stimulate circulation, relieve muscle aches and pains or merely to enjoy the aromatic experience. The concept, however, remains the same: imagine that you're relaxing in a giant cup of herbal tea.

There are several ways of brewing an herbal bath. The easiest is to toss several herbal tea bags directly into the bath water. Another option is to use pieces of muslin cloth which are filled with herbs and tied. These can be attached directly to the faucet so that the running water goes through them, or placed in the tub. Use the latter method with any of the following mixtures:

# Soothing Linden
# Bath Blend

## Recipe:

1/4 cup dried linden for soothing frayed nerves
1/4 cup dried comfrey for healing skin irritations
1/4 cup dried yarrow, an ancient
Chinese herb valued for its astringent as well as
muscle-soothing properties
2 tablespoons oatmeal to soften the water

# Revitalizing Ginger, Lemon & Parsley Bath Combination

## Recipe:
—

1/4 minced ginger root for promoting circulation
1/4 cup dried parsley, an herb valued by the
ancient Romans as a tonic for the skin
1/2 cup lemon peel (grated) for a cleansing
and aromatic effect
2 tablespoons oatmeal to soften the water

# Tonic Lemon Orange Bath

## Recipe:

—

1/4 cup grated lemon peel to treat lethargy
1/4 cup grated orange peel to
fight depression and anxiety
1 tablespoon dried parsley for stimulation
1 tablespoon dried comfrey, a mild antiseptic

# Invigorating Rosemary & Sage Bath

### Recipe:

—

1/4 cup dried rosemary to relieve mental fatigue
1/4 cup dried sage to treat loss of concentration
2 tablespoons oatmeal to soften the water

# Stimulating Basil, Eucalyptus & Peppermint Bath

### Recipe:

—

1/4 cup dried basil to regenerate mental powers
1/4 cup dried eucalyptus for treating
lack of concentration
1/4 cup dried peppermint to alleviate mental
fatigue and lack of concentration

# Essential Oil Baths

T he following recipes involve the use of essential oils (see list on pages 163-165) as a way to tone skin and to help remedy recurring problems. In order to avoid ending up with a film of oil on the surface of the bath water, it is best to add one capful of pure castile soap or very mild shampoo to disperse the oils in the bath water.

## *Hypnotic Neroli Oil Bath*

Neroli is an essential oil extracted from the fragrant flower of the orange tree, *Citrus aurantium*. It is named after *Anna-Maria de la Tremoille*, Princess of Nerola, a town near Rome. The oil's main properties are antispasmodic, sedative, tranquilizing, and slightly hypnotic. For insomniacs, neroli is a natural tranquilizer.

### *Recipe:*

—

10 drops neroli essential oil

*Drop oil directly into warm bath water and relax in it for 20 minutes.*

•

*For an energizing morning bath, add 5 drops of rosemary oil and 2 drops of lavender oil to bath water.*

*"In life, as in art, the beautiful moves in curves."*
—EDWARD GEORGE BULWER-LYTTON

# Thyme & Eucalyptus Back Relief Soak

Thyme is said to have been used for its medicinal properties as early as 3,500 BC. Throughout history, thyme (whose name comes from "thúmos," the Greek word for "smell") has been used to relieve asthma, coughs and colds; to kill yellow fever organisms; to reduce swelling provoked by gout; and to treat of nervous disorders. The Romans, who believed that thyme promoted bravery, would frequently bathe in the herb before going to battle. The following recipe is recommended for anyone who suffers from joint pains or backache. The eucalyptus oil enhances the action of the thyme.

## Recipe:

—

15 drops thyme essential oil
4 drops eucalyptus essential oil

*Mix oils in hot bath water and
soak for 15 minutes. After the bath,
massage joints and back with the
Thyme & Cedarwood Joint
and Backache Rub on page 61.*

JUNIPE

# Calming Juniper & Geranium Rose Bath

While this soothing and sedative, yet spiritually uplifting, treatment takes mere seconds to prepare, its calming and toning effects are long-lasting. Juniper, which has an extensive history of use, possesses many healing properties impacting the immune system, the circulatory system, the digestive tract and the skin. In the context of a bath, juniper oil will help combat mental fatigue and restore calm. It blends well with geranium oil, which is thought to stimulate the adrenal glands and balance hormones.

## Recipe:

---

6 drops juniper essential oil
6 drops geranium essential oil

*Add the essential oils to a bath
filled with warm water.
Relax in it for 15 to 20 minutes.
Enjoy.*

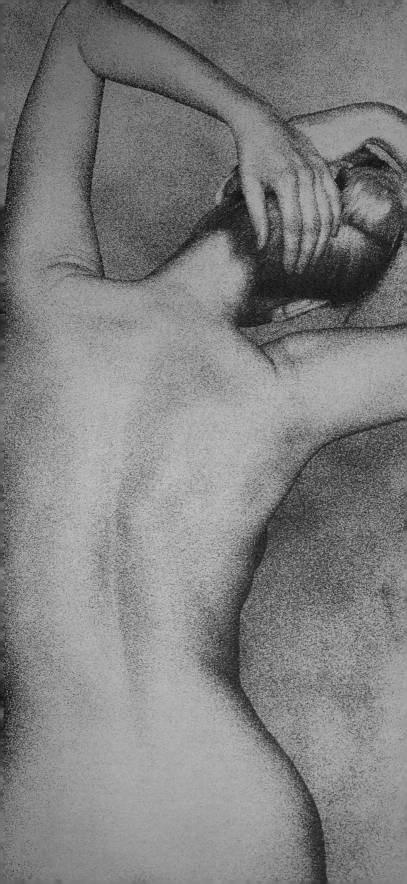

# Aphrodisiac Bath Blend

To relax the mind and stimulate sexuality and euphoria.

### Recipe:

—

4 drops ylang-ylang essential oil
3 drops neroli essential oil
2 drops bergamot essential oil
2 tablespoons honey
2 ounces fresh cream

*Combine ingredients in warm bath water and soak until euphoric.*

# Cypress & Orange Oil Bath for Oily Skin

Cypress oil, known for alleviating circulatory problems as well as menopausal symptoms, will help remedy oily skin when used in the bath. A touch of orange essential oil, a tonic for the muscular system, will help rejuvenate the skin, combat wrinkles and give the bath a fresh citrus aroma.

### Recipe:

—

4 drops cypress essential oil
4 drops orange essential oil

*Combine oils in a running bath and soak for 20 minutes.*

# Sandalwood & Sweet Almond Bath for Dry Skin

The Egyptians used sandalwood in medicine, embalming and religious rituals; the French used it to fight chronic bronchitis and urinary problems in the late 1800s. Today sandalwood is used primarily in perfume and soap making but has excellent skin healing properties, especially for cracked and chapped skin. Combined with sweet almond oil, a light but easily absorbed oil, it makes for a healing skin treatment.

## Recipe:

3 tablespoons sweet almond oil
8 drops of sandalwood essential oil

*Combine oils in warm bath water and relax in tub for 15 minutes. Pat dry and moisturize. This treatment should be followed every day until skin is healed.*

# Geranium & Olive Oil Bath Blend For Normal Skin

If you were being sent to a deserted island and were allowed to take only one essential oil with you, geranium oil should be your choice. Its multi-purpose characteristics include that of an insect repellent, tonic and antiseptic, mood lifter, athlete's foot remedy and frostbite healer. In the bath, geranium oil is particularly soothing for normal skin and has a wonderfully rich and relaxing aroma. Olive oil, one of the first oils ever used for cosmetic purposes, is high in vitamins and minerals and helps maintain the correct water balance and level of acidity for the body, both internally and externally.

### Recipe:

—

3 tablespoons olive oil
8 drops geranium essential oil

*Combine oils in warm bath water and
relax in it for 15 minutes.
Pat dry and moisturize as usual.*

•

*There are over 300 known essential oils
in the world. Ten of these may serve
together as a basic survival kit to soothe
ailments from the common cold to black
eyes to heart palpitations. These oils
are: lavender, eucalyptus, geranium,
rosemary, thyme, lemon and clove.*

# Seaweed Powder & Aloe Hydrotherapy Bath

This is the perfect thalassotherapy, or seawater therapy treatment, to concoct at home because it takes virtually seconds of preparation time and it truly is one of the best ways to unwind in the tub. Seaweed, or *Digita laminaria*, is very rich in iodine and protein. These nutrients are directly absorbed by the pores of the skin, affecting the body on a cellular level, and in the process will help detoxify, revive, remineralize and stimulate the body as a whole. Aloe complements the effect of the seaweed in that the gel helps regenerate and soothe damaged or dry skin. Next to fresh seaweed, the freeze-dried powder is seaweed in its purest form.

## Recipe:

—

1/2 cup freeze-dried seaweed powder
1/4 cup aloe gel

*Pour seaweed powder and aloe gel into bath water and relax for 20 minutes. Pat dry and moisturize. As seaweed baths can be draining, it is best not to overexert yourself immediately after the treatment.*

•

*"'Tis not a lip, or eye, we beauty call, but the joint force and full result of all."*

—POPE ALEXANDER

# Citrus & Milk Bath

Cleopatra was famous for bathing in milk and soothing her skin with milk's natural beauty-enhancing properties. Rich in protein, calcium and vitamins, milk is easily absorbed by the skin, leaving it smooth and moisturized. It also helps ease nervous tension and restlessness. The lavender and citrus peel in this bath formula stimulate the circulatory system and add a wonderfully luxuriant aroma to the treatment.

### Recipe:

—

1 cup dried milk powder
*(if you have oily skin, use nonfat milk)*
1/4 cup orange peels
1/4 cup lemon peels
4 drops lavender essential oil

*Draw a warm bath and while water
is flowing slowly add dried milk, orange
and lemon peels and lavender oil.
Soak for 20 minutes.
Pat dry and moisturize.*

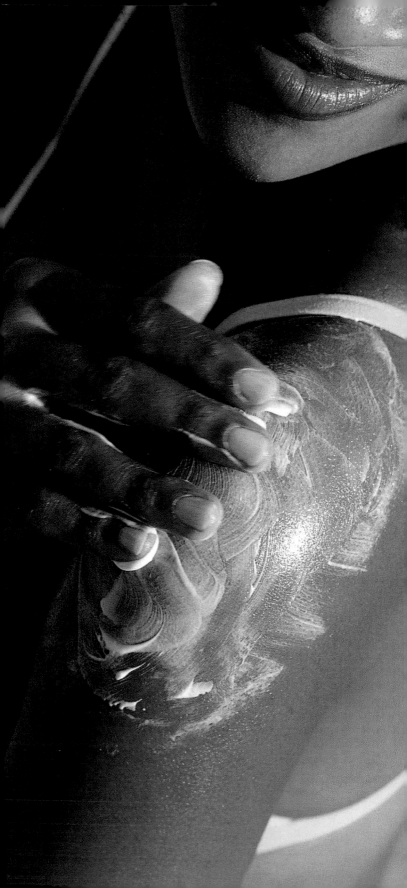

## *Apply moisturizer every day after bath or shower.*

- It's best to apply moisturizer when the skin is still warm and moist.

- Make sure you choose a moisturizer with ingredients that are suitable to your skin type. The trick is to find one that does not feel greasy and will not clog pores.

- Rich, heavy oils such as olive, sesame and safflower are best for dry skin. Lighter oils such as jojoba, avocado and grapeseed are finer and more suitable for normal to oily skin.

- Gels, such as aloe, are predominantly water-based and suitable for normal to oily skin.

- Stay away from mineral oils (petroleum-based) such as "baby oil" and oils that contain animal fat. These oils are not absorbed by the skin and tend to form a greasy layer on the skin's surface that can cause irritation.

> *"The ideal of beauty is simplicity and tranquillity."*
>
> —Johann Wolfgang von Goethe

# Coconut Body Butter for Dry Skin

Coconut oil, high in vitamins, minerals and saturated fats, is a wonderful skin softener. Cocoa butter, which is solid at room temperature, is obtained from seeds of the cacao plant. Together, these ingredients preserve and effectively protect the skin from the wind and cold weather by coating the skin's surface with a thin layer that locks in the skin's natural moisture. In addition, sesame oil is one of the few natural oils that contains sun-screening properties. Combined with olive oil, this body butter is ideal for smoothing on skin that has been over-exposed to nature's elements.

## Recipe:

—

2 tablespoons grated beeswax
2 teaspoons distilled water
1/2 cup cocoa butter
3 tablespoons sesame oil
2 tablespoons coconut oil
1 tablespoon olive oil

*Melt the beeswax over low heat with the water. Spoon in cocoa butter and blend. Gradually blend in sesame oil, coconut oil and olive oil. Pour into a glass jar. Lotion will thicken as it cools. Apply liberally over body.*

# Rose & Galbanum Body Softener for Normal to Oily Skin

Galbanum is a resinous gum or sap obtained from the Ferula plant, native to southern Europe and parts of Africa and Asia. When ingested, galbanum has sedative and anti-spasmodic properties. When applied to the skin, it alleviates skin irritations and is a particularly effective skin regenerator. The oil is thick and has a slight earthy smell to it. Combined with rose oil, which fights wrinkles and broken capillaries, this body lotion can also be used for the face and hands.

## Recipe:

2 tablespoons grated beeswax
1/4 cup distilled water
1/2 cup almond or apricot oil
1/4 cup rose water
5 drops galbanum essential oil
4 drops rose essential oil

*Melt the beeswax on low heat with the distilled water. When it is melted, pour into a blender and blend on low speed. While blender is running, gradually pour in almond or apricot oil, rose water and essential oils. The mixture will have the texture of frosting and will thicken as it cools. Pour the lotion into a glass jar or container. As this is a very rich and thick lotion, use it sparingly. Its shelf life is approximately 3 to 5 months.*

## Cooling Witch Hazel & Wheat Germ Oil Moisturizer

Witch hazel has very cooling, styptic and cleansing properties and is often used in after-shaves and skin toners. Wheat germ oil is high in vitamin E and is an effective antioxidant. The lemon oil adds a cool refreshing scent.

### Recipe:

—

2 tablespoons grated beeswax
2 tablespoons distilled water
1/4 cup wheat germ oil
2 tablespoons witch hazel
2 drops essential lemon oil

*Melt the beeswax and distilled water over low heat then pour it into a blender. On low speed, gradually add wheat germ oil and witch hazel and blend until smooth. Add essential oil and pour into a jar. Lotion will thicken as it cools. Because of its cooling effects, this moisturizer is ideal for hot summer months.*

*"Grace has been defined as the outward expression of the inward harmony of the soul."*

—WILLIAM HAZLITT

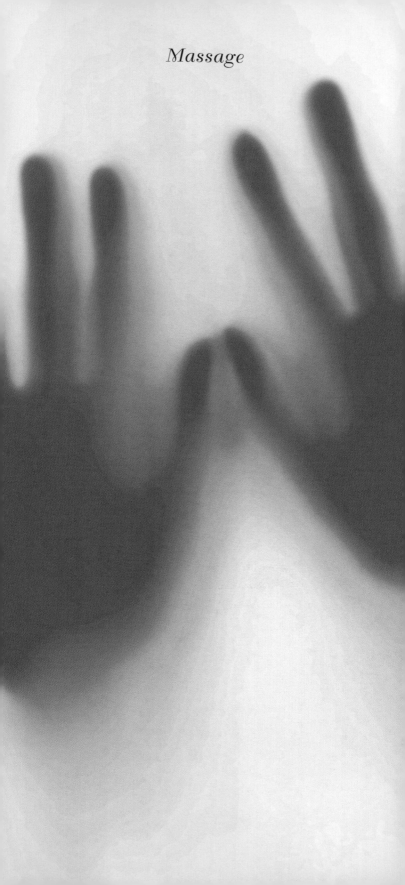

*Massage*

# Get a partner.

- If you don't have a massage table, a firm mattress covered with a bath sheet or towel will work.

- Select calming music and light a few aromatherapy candles.

- The primary purpose of a massage is to activate the body's nerve endings and stimulate the circulation of the blood through gentle, and sometimes rougher, stroking of the skin.

- There are many different massage techniques. Kneading, deep friction, rocking, compression, skin rolling, fingertip brushing and percussion are just a few. The type of technique used will depend on what works best for your body and what feels most comfortable. A massage that is too painful will not yield effective results.

- Essential oils used in massage treatments can relieve stress and anxiety, increase lymph production and drainage, improve blood flow, detoxify, ease pain and soothe sore muscles. Custom blend your oil to meet your specific needs.

- Warmth is an important factor in massage therapy as it keeps the muscles relaxed. Make sure the room stays warm and use a space heater if available. Keep a few blankets nearby for after the massage.

- Pour massage oil first into your palm, instead of directly on the skin, to warm the oil.

- After the massage treatment, it is important to stay warm and relaxed for at least 20 minutes.

- Replenish the body's fluids by drinking water or herbal tea.

- It is not advised to get a massage immediately after exercising, as the body is still in the process of perspiring and eliminating toxins, which will interfere with the skin's absorption of the oil.

"*The human body
is a machine which winds
its own springs.*"

—JULIAN OFFROY DE LA METTRIE

# Lavender & Clary Sage Blend Massage Oil

Grapeseed oil is high in polyunsaturates, extremely light and easily absorbed by the skin. It works well with essential oils by facilitating their rapid penetration into the body through the pores of the skin. Clary sage oil is pale yellow-green and helps alleviate fatigue, irritability and depression. Lavender, which is extremely aromatic, is known for its headache/stress relieving and skin-soothing properties and stimulates the circulatory system. This lavender and clary sage blend is ideal for cleansing and ridding the body of toxins and impurities and for total physical and spiritual relaxation.

### Recipe:

—

1/2 cup grapeseed oil
4 drops lavender essential oil
4 drops clary sage essential oil

*Blend oils together
and massage into skin.*

# Vetiver & Jojoba Massage Oil

Jojoba oil, which is actually more like a wax than an oil, is one of the most popular carrier oils. Its cosmetic properties are similar to those of natural human oil secretions (sebum) and it is suitable for all skin types. Rich in vitamin E, jojoba is easily absorbed by the skin and has the added advantages of having a longer shelf life than most carrier oils and blending well with just about any of the essential oils. Vetiver essential oil, chosen here for its calming and grounding properties in addition to its soothingly rich and woody scent, can be substituted for a number of essential oils (see pages 163-165), depending on what most suits your mood.

### Recipe:

—

1/2 cup jojoba oil
40 drops vetiver essential oil

*Mix the oils and store in an amber glass
bottle. Use for full body aromatherapy
massages or for local applications
to soothe tired and sore muscles.*

### "Exuberance is beauty."
—WILLIAM BLAKE

# Thyme & Cedarwood
# Joint and Backache Rub

Rich in thymol, carcavol, menthone and pinene, this thyme rub is ideal for the relief of backache and joint pains and should be used immediately after the bath or shower Cedarwood oil enhances the healing effects of thyme and adds a nice woody scent to the mixture.

### Recipe:

—

2 ounces jojoba oil
12 drops of thyme essential oil
12 drops of cedarwood essential oil

*Mix ingredients thoroughly and
massage into affected areas
with a slow circular motion.*

# CHAPTER 2

# *Face*

CLEANSERS

·

STEAMS

·

MASKS AND PEELS

·

TONERS

·

MOISTURIZERS

·

# Cleansers

The notion of classical beauty has changed over the centuries and varied from culture to culture. The ancient Greeks believed harmonious facial proportions were the key to a woman's beauty. Victorians favored thin rosebud lips, quite contrary to the lush and full lips of today's magazine beauties. And Rubens portrayed his goddesses with plump cheeks and double chins.

Achieving a beautiful look by society's standards, women have covered their faces with substances made of lead, as the elegant Elizabethan women did to achieve a ghost-like pallor, and have smothered mercury-based lotions to get rid of freckles and spots. They have also smeared wax on their wrinkles, as did many women in ancient Rome.

Regardless of what constitutes the "ideal face," what all beautiful women have in common today is the look of clean, healthy and toned skin. Whether it's the ravages of time, too many late nights, nature's elements or everyday stress, the face's sensitive skin is constantly under attack and needs to be pampered on a regular basis.

While you can always camouflage hands in gloves, feet in shoes, a body in clothes and hair in a hat, your naked face remains your signature. The time you spend nurturing it will certainly help you put your best face forward.

- When applying a treatment to your face, don't forget your neck. The skin on your neck ages the same way as the skin on your face does.

- Always clean your face before doing a steam.

- Steam your skin once a week to clean pores, to increase circulation in the facial capillaries and to moisturize.

- When doing a facial steam, be careful not to scald your skin with the steam. If it is burning the inside of your nose, then wait a few minutes for the water to cool down before you try again.

- When applying masks or peels, use dark-colored towels, as some of the ingredients might stain.

- Pull your hair back so you can get right up to the hairline.

- Always use sunscreen when outside.

- Vitamin A lubricates and heals the skin. Try to add more to your diet by eating foods rich in vitamin A such as spinach, cantaloupe, carrots and pumpkin.

- Vitamin E promotes skin elasticity so try to add whole grains, eggs, leafy greens, and broccoli to your diet.

- Always moisturize area above upper lip, and below eyes but above cheekbones as these particular areas tend to lack moisture.

## *The five key steps to the perfect face:*

1. CLEANSE: *to rid the skin of dirt, makeup and oils.*
2. STEAM: *to open the pores, increase blood circulation and deep-cleanse the skin.*
3. MASK OR PEEL: *to further remove residue deep within the pores and to nourish and replenish the skin with essential vitamins and minerals.*
4. TONE: *to tighten the pores and prepare the skin for moisturizing.*
5. MOISTURIZE: *to replenish the skin with fluid and finely coat it with a protective film.*

# Thyme & Fennel Seed Cleanser For Normal Skin

Fennel, or *Foeniculum vulgare*, has been used throughout history as an aid to digestion or as a slight diuretic. As an infusion, fennel seeds can be gently cleansing and toning for the skin, and they can help reduce puffiness and superficial irritation. Thyme, which is used in antiseptic preparations, is a good astringent. Because this cleanser is so gentle, it can be used every morning. Simply dab it on face and neck with a cotton ball.

### Recipe:

—

2 sprigs fresh thyme, crushed
(or 1/2 tablespoon dried thyme)
2 teaspoons fennel seeds, crushed
1/2 cup boiling water
juice of half a lemon

*Mix thyme and fennel seed in a bowl and cover with boiling water. Add lemon juice and let steep for 15 minutes. Strain infusion and store liquid in a jar. Keep in the refrigerator.*

•

*Slice a ripe tomato and rub it all over your face. Let the juice soak in for about 5 minutes. The tomato's mild acid will bring dirt and other impurities to the surface of the skin. Rinse thoroughly with water.*

# Apple Cider Wash for Oily Skin

If you don't have much time on your hands, this wash is very effective for removing dirt and residue from face.

### Recipe:
—

2 tablespoons apple cider vinegar
2 tablespoons distilled water

*Mix water and apple cider vinegar
in a small bowl and apply to face
with a cotton ball. Avoid eye area.
Rinse with tepid water. Pat dry.*

•

*Crisco is an excellent make-up remover.*

## Cleansing Oatmeal & Honey Paste

Ground oatmeal is a gentle exfoliator that removes dead surface cells and residue. It can be used as a mild cleanser instead of soap, and helps restore the skin's moisture. The combination of yogurt, which will soften the skin, and honey, a natural humectant, makes for an effective, everyday cleansing moisturizer.

### Recipe:

—

1/2 cup oatmeal
2 tablespoons honey
1/4 cup plain yogurt or buttermilk

*Finely grind oatmeal in a blender or food processor. In a small bowl, combine honey, yogurt and add ground oatmeal. Mix thoroughly until it has a paste-like consistency. Smooth over face and neck and leave on for 15 minutes. Rinse off with warm water and pat dry. Can be applied every day, preferably in the morning.*

•

*When doing facial treatments, don't forget the neck—it needs as much pampering as the face.*

# Rose Petal Facial Steam for Normal Skin

This recipe is similar to the one for Rosewater (see page 94), except that this one requires more water. After you're done with the steam, you can easily make a batch of rosewater with the leftover ingredients.

## Recipe:

—

1 cup fresh rose petals
1 gallon almost boiling water

Put fresh rose petals in a large bowl and
pour almost boiling water over them.
Place towel over head and lean face over bowl,
about 12 inches away, to absorb steam and
vapors. Keep breathing in vapors for
15 minutes. Gently pat dry.

Once you're done with the steam, you can
make rosewater. First, strain the steam liquid
and save the petals and the water separately.
Pour water into a medium saucepan and
simmer for about 30 minutes, or until there is
approximately one cup of liquid left.
Once reduced, pour it over the saved petals
and steep for another 20 minutes. Strain the
liquid again. Discard petals and pour liquid
into a glass bottle and allow to cool.
You have just made rosewater!

# Peppermint and Rosemary Facial Steam

This herbal steam will open the pores, extract impurities and clear sinuses. It's also very helpful in alleviating headaches.

### Recipe:

—

4 drops peppermint essential oil
4 drops rosemary essential oil
1 gallon boiling water

*Pour boiling water in a large bowl and let cool slightly before adding the essential oil. Being careful not to burn yourself with the steam, put face over bowl and cover your head with a towel, allowing vapors and steam to accumulate for 10 minutes.*

•

*Skin is the largest organ of the body and is subject to a lot of stress.*

# Lavender Facial Sauna for Oily Skin

Lavender normalizes the secretion of sebaceous glands and is most beneficial for acne-prone skin.

### Recipe:
—

6 drops lavender essential oil
1 gallon almost boiling water

*In a large bowl, add essential oil to almost boiling water. While being careful not to scald skin, hold face over bowl with a towel covering your head. Allow steam and vapors to accumulate for 10 minutes. Pat dry.*

# Bergamot Facial Sauna for Blemished Skin

Emerald green bergamot oil is used primarily for its antiseptic properties. In a facial steam, it will reduce redness, irritations and puffiness.

### Recipe:
—

12 drops bergamot essential oil
1 gallon almost boiling water

*Pour almost boiling water in a large bowl and add drops of bergamot essential oil. Place towel over head and hold face over the bowl about 12 inches from water. Blemished skin should steam for no longer than 5 minutes. Gently pat dry.*

QUICK SKIN-TIGHTENING TIP:

•

*Beat an egg white until it is frothy,*
*apply to face and leave on for 5 minutes.*
*The mask will tighten. Rinse off*
*thoroughly, first with warm water*
*then with cool.*

"*Beauty: the adjustment of all parts
proportionately so that one cannot add
or subtract or change without impairing
the harmony of the whole.*"

# Seaweed Detoxifying & Remineralizing Mask

When selecting a seaweed powder, make sure it is the kind that can be used on the face. Certain types of seaweed are meant for body wraps only and would be too harsh on the delicate skin of the face. Kelp, a seaweed which grows in rich ocean beds, is known to nourish the sensory nerves and brain membranes. It is also considered a natural emollient, and is very high in vitamins and minerals. While the mask will pull impurities out of the skin, the seaweed will help restore some of its lost nutrients.

## Recipe:

—

4 tablespoons kelp powder
1/2 cup aloe vera gel
3 tablespoons distilled water

Combine kelp powder and gel in a small bowl. Slowly add water until mixture reaches the consistency of a thick paste. Apply to face and neck with fingers and relax for 15 minutes. Rinse with tepid water and pat dry. Moisturize.

*"Though we travel around the world to find the beautiful, we must carry it with us or we find it not."*

—RALPH WALDO EMERSON

## Carrot, Avocado & Cream Nourishing Mask

The combination of carrot (high in beta-carotene and anti-oxidant vitamins), heavy cream (high in calcium and protein) and avocado (a rich source of vitamin E) in this facial mask will improve skin texture, diminish age spots and rebuild skin collagen when used with regularity.

*Recipe:*

———

1/2 cup heavy cream
1 carrot, cooked and mashed
1 avocado, peeled, pitted and mashed
3 tablespoons honey

*Combine ingredients in a bowl and spread over face and neck. Relax for 15 minutes. Rinse with cool water.*

•

*The thicker the clay of the mask, the more intense the action on the skin.*

•

*An avocado pit makes a natural massage tool: rub it and roll it against your skin to soothe tired muscles.*

# Pineapple & Olive Oil Mask

Pineapple contains bromelain, a protein-digestive enzyme which helps rid the skin of dead cells and dirt. In addition, it helps counteract histamines and has anti-inflammatory properties. The fruit is also a mild astringent and skin freshener. Olive oil, which has excellent healing properties and is a good source of vitamin E, will help restore the skin's surface. Make sure you use the highest quality cold-pressed extra-virgin oil.

## Recipe:

—

4 large pineapple chunks or
1/2 cup canned, drained pineapple
3 tablespoons olive oil

*Place ingredients in blender and blend
until almost smooth. Apply mixture
to face with fingers and leave on
for 15 minutes. Rinse face with warm
water and pat dry.*

"Life is not merely being alive,
but being well."

— ANONYMOUS

# Soothing Cucumber Mask

Cucumber is a cool and mild astringent, and is especially soothing for irritated skin. The chamomile and green tea will help alleviate puffiness; the aloe vera helps maintain the skin's moisture balance as well as helping heal sunburned skin. This mask can be used every morning or three times a day for sunburned skin.

## Recipe:

1 small cucumber, peeled and seeded
2 ounces green tea, steeped and strained
2 ounces chamomile tea, steeped and strained
1 packet unflavored gelatin
1 ounce aloe vera gel

*Place cucumber in blender and purée until smooth. Strain purée through sieve or coffee filter and reserve juice. In a small saucepan, combine green tea, chamomile tea and gelatin. Stir over low heat until dissolved. Remove from heat and pour into a glass bowl. Add cucumber juice and aloe gel to mixture. Place in refrigerator for approximately 25 minutes until the mixture has started to thicken. Spread over face and neck with fingers and allow to dry for 20 minutes. Peel off mask and rinse face and neck with warm water. Pat dry.*

## Weekly Rose Oil & Honey Beauty Mask for Maturing Skin

The origin of the rose has inspired many legends—a jealous goddess created the flower to rival Venus' beauty, the first rose was created from a drop of sweat falling from the prophet Mohammed's brow and Cupid offered a rose to the god of silence to bribe him not to reveal Venus' lovers. One thing remains certain: the rose, its steam-distilled oil and by-product (rosewater) have innumerable therapeutic properties, from treating respiratory coughs to curing ulcers. In beauty treatments, rose oil is particularly effective for treating wrinkles, puffiness and dry skin and serves as an excellent face and neck toner.

### Recipe:

—

2 tablespoons honey
2 teaspoons almond oil
5 drops rose essential oil

*Mix honey, almond oil and rose oil. Apply to face and neck with fingers. Relax for 15 minutes, then rinse off with warm water. Pat dry.*

•

*5,000 pounds of rose petals—as many as 100 million—are required to produce one pound of rose oil.*

# Banana & Chickpea Protein Mask

Chickpea flour has many advantages: it's an effective skin-softening exfoliator, it stimulates circulation and it rejuvenates tissues. Bananas, rich in potassium and vitamin A, are natural emollients and help reduce redness and puffiness. The egg in this mask will tighten pores and leave your skin feeling revitalized.

## Recipe:

—

4 tablespoons chickpea flour
1 ripe banana, peeled and mashed
1 egg, beaten

*In a bowl, blend chickpea flour and ripe
banana into a paste. Add beaten
egg and apply mixture to face and neck.
Leave on for 15 minutes and
rinse well with warm water. Pat dry.*

•

*Almond, soy, boraye, rosehip
and evening primrose oils have a
relatively short shelf-life and go
rancid quickly. To extend shelf-life,
it is best to blend them with
jojoba, oil, wheatgerm or olive oil.*

## Papaya Face Peel

Papaya contains papain, a natural enzyme that removes dead skin cells.

### Recipe:

—

1 packet unflavored gelatin
3 tablespoons distilled water
1 papaya, peeled and seeded

*In a saucepan, combine the gelatin and water and dissolve the gelatin over low heat. Place papaya in a blender and blend thoroughly. Strain and save the liquid. In a small bowl, combine gelatin and papaya juice. Refrigerate for about 20 minutes or until the gelatin is just beginning to set. Spread gelatin over face and neck and relax for 15 minutes. Rinse off with a soft sponge or cloth dipped in warm water.*

*"Remember that the most beautiful things in the world are the most useless; peacocks and lilies for instance."*

—JOHN RUSKIN

# Rosewater Astringent

Rosewater is a by-product of rose essential oil and can also be made by distilling fresh rose petals in water. Rosewater is a natural astringent and is easy to apply with cotton balls. Since certain roses are chemically treated to enhance color and prolong their longevity, make sure the petals you use are entirely chemical-free.

## Recipe:

—

1 cup fresh rose petals
1 cup distilled water

*Place fresh petals in a bowl. Bring water to a boil and pour over rose petals. Let steep for 30 minutes. Strain petals from liquid and store in a glass bottle in the refrigerator.*

# Green Tea Tonic

This simple tonic cleanses, tones and soothes. It is recommended for aging and tired skin.

## Recipe:

———

2 teaspoons powdered green tea
1/2 cup boiling water

*Steep green tea in boiling water for
10 minutes and allow to cool.
Apply tonic to face with cotton or
gauze. Use daily for best results.*

## Lemon Mint Toner

This toner is ideal for normal or slightly oily skin. It will remove any residue, close pores and restore balance to the skin.

### Recipe:

—

1 peppermint tea bag
1 cup boiling water
1/4 cup witch hazel
1 tablespoon lemon juice

*Place tea bag in boiling water and steep for 15 minutes. Discard tea bag and allow liquid to cool. Add witch hazel and lemon juice and store in a glass container in the refrigerator. Apply daily after cleansing the face.*

### QUICK-FIX EYE SOOTHER

•

*You wake up in the morning, your eyes are puffy. Before getting in the shower, place two teaspoons or 2 slices of cucumber in the freezer. After you've finished with your shower and have gotten dressed, place the frosted spoons or the cucumber slices on your eyelids and relax for 10 minutes.*

# Cranberry & Savory Acne Inhibitor

The acidity in the cranberries works as a mild astringent and toner. Savory, which the Greeks originally considered to be an aphrodisiac, and which possesses antiseptic properties similar to those found in oregano and thyme, helps cleanse the skin of excess oils.

## Recipe:

1 tablespoon dried savory leaves, crushed
1/2 cup boiling water
1/2 cup cranberries

*Place dried savory in a small bowl and cover with boiling water. Allow to steep for 15 minutes, then strain liquid and set aside. Place cranberries in a blender and purée. Strain pulp out of liquid, discard pulp and combine the liquid with the savory infusion. Soak cotton pads or a washcloth in the liquid and leave on face for 15 minutes. Try to avoid the eye area. Remove pads or washcloth and rinse with warm water.*

# Rich Coconut Oil Night Cream for Dry Skin

This night cream is rich in lecithin (also present in the egg yolk), which helps rebuild and regenerate cell tissue. Combined with the sun-screening properties of sesame oil and lemon's citric acid, which kills bacteria on the skin, the coconut oil will help seal in the skin's moisture and leave your face feeling incredibly smooth.

## Recipe:

—

1 egg
1/4 cup sesame oil
1/4 cup sunflower oil
1 tablespoon liquid lecithin
1 tablespoon fresh lemon juice
1/2 cup coconut oil

*At low speed, blend egg and sesame oil in a blender. While blender is still running, slowly add sunflower oil, lecithin and lemon juice. Gradually add coconut oil to the mixture and blend until mixture is thick. Refrigerate for one hour and then blend again at low speed. Apply to face and neck before going to bed. This night cream should be stored in a jar in the refrigerator and may be kept for up to 2 weeks.*

# Cucumber Cold Cream

This particular cold cream is very refreshing. Cucumber juice, a mild astringent, is perfect for cleansing sensitive skin. Lanolin, which is more like a wax than an oil, absorbs and holds water next to the skin.

*Recipe:*

—

2 small cucumbers, peeled and halved
1 cup lanolin
1/4 cup almond oil
1/2 cup boiling water

*Place cucumber in a blender and blend thoroughly until liquefied. Slowly add boiling water to cucumber liquid and set aside to cool. In a saucepan on low heat, melt lanolin. Remove from heat and, stirring constantly, gradually add almond oil. Strain cucumber pulp through cheese cloth or strainer and reserve juice. Gradually add cucumber juice to lanolin and oil mixture while continuing to stir. Store cream in a small glass or plastic jar and keep in the refrigerator.*

# Aloe & Rosewater Moisturizer

Rosewater, a by-product of rose essential oil, has anti-inflammatory, antiseptic and toning qualities. Combined with aloe gel, this lotion will leave your skin supple and moist.

## Recipe:

—

1/2 cup olive oil
2 tablespoons aloe vera gel
3 tablespoons grated beeswax
3 tablespoons lanolin
4 tablespoons rosewater (to make your own rosewater, see recipe page 94)

*In a small bowl, blend olive oil and aloe gel and set aside. In a small saucepan, melt beeswax and lanolin over low heat, stirring constantly. Remove from heat. Gradually pour in aloe and olive oil mixture while continuing to stir. Add rosewater and transfer the mixture to a blender and blend until very smooth. Pour into a jar and refrigerate. Moisturizer will solidify as it cools.*

# Sweet Orange Wrinkle Banisher

Particularly effective against wrinkles caused by over-exposure to the sun, orange oil helps rejuvenate and revitalize the skin. The almond oil will help replenish the skin's moisture.

### Recipe:

—

2 teaspoons hazelnut oil
1 teaspoon almond oil
8 drops sweet orange essential oil

*Combine oils and massage gently into the skin. Use morning and night after cleansing.*

*"Our body is a machine for living. It is organized for that, it is its nature. Let life go on in it unhindered and let it defend itself, it will do more than if you paralyze it by encumbering it with remedies."*

—LEO TOLSTOY

# CHAPTER 3

# Hands & Feet

CUTICLES AND NAILS

•

HAND AND FEET SCRUBS, EXFOLIATORS AND WASHES

•

MOISTURIZERS

•

# Cuticles & Nails

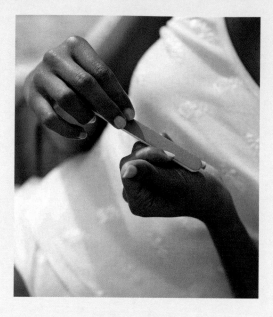

If the eyes are the window to the soul, then the hands are the keepers of time. Hands are generally prone to age at a much faster rate than the other parts of the body. They are often the first to express signs of everyday wear and tear, in part because of the relentless exposure to soaps, detergents, cold, heat, wind and water. In order to minimize the harsh effects of these elements it is crucial to keep replenishing the skin's surface with moisture and nutrients, both from inside and outside. Because we rely so heavily on the use of our palms, fingers and nails during the course of the day, an optimal time for a hand treatment—with the exception of a massage, perhaps—is at night.

While it is true that feet are not prone to the same kinds of stresses as are hands by virtue of the fact that shoes and socks generally protect them, they are nevertheless subjected to a different kind of abuse: hard pavements, high heels, excessive body weight and poor circulation. Although few treatments can rival the effects of foot shiatsu, a warm soak in mineral-rich salts will certainly help dissolve the aches and restore the senses.

# To keep fingernails and cuticles healthy, eat a healthy diet, rich in vitamins.

- Try not to use your nails as a tool.

- Don't bite your nails or cuticles.

- Never cut your cuticles or use cuticle removers. Soak your nails in warm water to soften cuticles, then gently push back the cuticle using a soft, moist towel. Only when necessary, trim excess dead skin with small, sharp scissors.

- Use a nail brush with baking soda, not a sharp object, to remove dirt from underneath nails.

- Give yourself—or better yet, have someone give you—a manicure or pedicure at least once a week.

- Avoid frequent use of nail polish remover, which dries out the nail bed and causes it to split.

- Avoid quick-dry nail polishes generally high in acetone, as they will dry the nail.

- If you're going to buff nails, do not apply heavy pressure, as this may thin the nail.

- As with hands, keep nails and cuticles away from drying soaps, detergents, and hot water as much as possible.

- Massage and moisturize nails and cuticles as often as possible.

*To brighten yellowed and dull nails, scrub them with white vinegar.*

# Almond-Jojoba Nail & Cuticle Treatment

The oil in this treatment locks in moisture and the honey conditions and strengthens nails.

### Recipe:
—

2 tablespoons almond oil
2 tablespoons jojoba oil
2 tablespoons olive oil
2 teaspoons honey
1 teaspoon vitamin E oil or
3–4 vitamin E capsules, broken open

*Mix together all the ingredients. Soak nails in warm water for 10 minutes and pat dry. Massage treatment lotion into hands and feet, concentrating on nails and cuticles. Repeat after washing hands and feet, or after a bath or shower. Recipe yields enough for 4 hand treatments, or 2 hand treatments and 2 feet treatments.*

# Nail Whitener

When used twice a week, this soak will whiten stained and dull nails.

### Recipe:
—

1 tablespoon hydrogen peroxide
1 cup of warm water

*Soak nails in solution for 15 minutes. Pat dry.*

# Hot Oil Nail Treatment

Olive oil will promote nail strength and flexibility.
Mixed with vitamin E, the oil nourishes cuticles and nails.

## Recipe:

—

1/4 cup olive oil
oil from 1 vitamin E capsule

*Heat oil in saucepan and let cool until just
warm to the touch. Add vitamin E oil
and rub into cuticles and nails, massaging
any excess oil into hands.*

# Walnut Nail-Hardening Oil & Hand/Foot Scrub

Castor oil is particularly beneficial for nails, hair and lips. When combined with olive oil, it enhances the nails' flexibility and renders them more pliable and less prone to chipping and breaking. The ground walnuts provide the appropriate friction for the removal of dead surface cells, while replenishing the skin's moisture. Honey is a natural antiseptic and bacteria inhibitor. It is also high in potassium and is ideal for any hand treatment.

## Recipe:

—

1/2 cup shelled walnuts
1 tablespoon olive oil
1 tablespoon castor oil
1 teaspoon honey

*Grind walnuts to a coarse powder.
Add olive oil, castor oil and honey to make a
thick paste. Vigorously rub hands and feet
with the paste. Rinse with warm water.
Pat dry and moisturize. This treatment is
most effective when used twice a week.*

# Hand & Foot
## Scrubs, Exfoliators and Washes

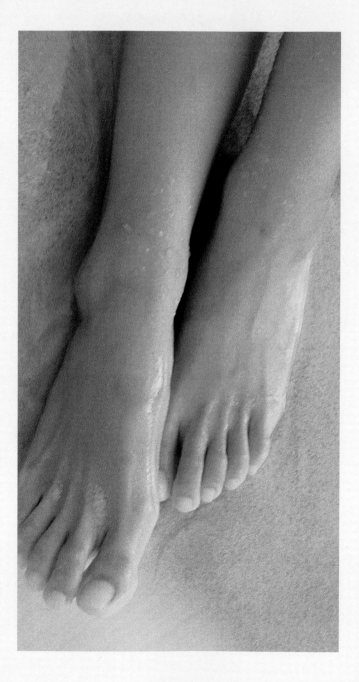

# Massage hands and feet once a day using your fingers or an avocado pit.

- Wear rubber gloves when using soaps, detergents or chemicals,to avoid rashes, use gloves lined with natural fibers as opposed to those with synthetic liners.

- Choose mild ingredients (such as milk and cream) for your hand cleansers.

- Use gloves when you peel citrus fruit.

- Before you go out in the cold, moisturize and put gloves on immediately afterwards for an instant nourishing treatment.

- Clean your rings frequently by soaking them in a solution of white vinegar and distilled water.

- Use a hand mask once a week.

- For deep moisturizing, apply a rich moisturizer, wrap hands with plastic wrap and leave on for 30 minutes; or put cotton gloves on over moisturizer before going to bed.

- Soak feet for 20 minutes in a warm footbath before exfoliating. Add a couple of drops of peppermint essential oil for stimulation.

- If you're going to treat your hands, don't forget your feet!

- Wear cotton socks that keep your feet nice and fresh as opposed to nylon socks which trap heat and moisture.

- Take long walks barefoot on a sandy beach. Sand is a natural exfoliator and its friction on the skin is an ideal massage.

- Avoid shoes that are too tight.

- It's best to buy shoes in the middle of the day as feet will have had a chance to swell up a bit. Sizing a slightly swollen foot will ensure a comfortable fit.

- Drink plenty of water to replenish the body's fluids.

# Apricot Oil & Sugar Hand Exfoliator

Sugar abrasion is a popular method for cleaning and stimulating the skin. The apricot oil serves as a natural lubricant, moisturizer and skin rejuvenator.

## Recipe:
—

1/2 cup granulated sugar
2 tablespoons apricot oil
juice of 1/2 lemon

*Combine ingredients and*
*immediately rub the mixture on hands.*
*Rinse with warm water.*
*Pat dry and moisturize.*

# Cornmeal & Pumice Foot Scrub

Cornmeal is a mild abrasive and, when used in conjunction with a pumice stone, exfoliates and smoothes rough, dry skin.

## Recipe:

—

1/2 cup dry cornmeal
2 tablespoons avocado oil
1 pumice stone

*Mix cornmeal and avocado oil. Spread on feet and rub skin with a pumice stone, concentrating on heels and callused areas. Rinse with warm water and pat dry.*

# Cooling Cucumber Hand Wash

This mild wash works well for irritated skin. Cucumber juice is a mild astringent and relieves puffiness and helps heal abrasions.

### Recipe:

—

1 small cucumber, peeled
1 tablespoon witch hazel

*Place cucumber in a blender and blend until liquefied. Add witch hazel to cucumber juice and wash hands with liquid. Pat dry and moisturize.*

# Healing Chickweed & Lavender Hand Treatment

Fresh chickweed possesses certain healing qualities for cuts and irritations. Combined with lavender oil, this treatment works well on paper cuts and itchy hands.

## Recipe:

—

1 cup fresh chickweed, crushed
1 cup grapeseed oil
30 drops lavender essential oil

Place crushed chickweed in a glass jar and cover with grapeseed oil. Store in a glass container in a dark place for 3 weeks. After 3 weeks, strain oil and discard chickweed. Add lavender essential oil. Apply to cuts or affected areas with cotton as often as necessary.

"She got her good looks from her father—
he's a plastic surgeon."

—Groucho Marx

# *Lavender Flower Hand & Foot Wash*

Use this wash to relieve red and inflamed hands and feet.

## *Recipe:*

1/2 cup dried lavender flowers
1/2 cup fresh sage, finely chopped
2 cups water
8 drops lavender oil

*In a saucepan, combine lavender flowers, sage and water and simmer, covered, on low heat for about 20 minutes. Strain mixture through cheese cloth and let cool. Discard solids. Add lavender oil and pat on hands and feet with a wash cloth. Repeat as necessary.*

# Avocado Hand Rub

To exfoliate, deep cleanse and moisturize the hands.

### Recipe:

—

1 ripe avocado, peeled and pitted
3/4 cup oatmeal
1/4 cup almond oil
1/4 cup olive oil
1/4 cup water

Place avocado meat, oatmeal, almond oil
and olive oil in blender and blend until smooth.
At low speed, drizzle in water until paste
has the consistency of thick pudding. You may
not need all of the water. Rub and massage into
hands in a circular motion. Rinse off with
tepid water and pat dry. Use twice a
week to maintain results.

# Deodorant Green Clay Foot Mask

Clay is ideal for pulling impurities from the system and restoring minerals directly into the skin while gently exfoliating at the same time. It also serves as an effective heat conductor, which makes for a soothing and muscle-relaxing treatment for tired, aching feet. There are many different kinds of clay, varying in color (from green, to red, to pink to white), intensity, purpose and origin. Green clay, however, works best for this particular treatment. It is rich in magnesium and silica, highly absorbent and makes a good foot deodorant.

## Recipe:
—

1/2 cup powdered green clay
1/2 cup distilled water

*Prepare a thick paste with the clay and the water. Coat feet with clay and allow to dry. Rinse off with warm soapy water and pat dry. Moisturize if necessary.*

## Herbal Foot Soaks

Simple to prepare, herbal soaks are ideal for tired feet. Chose from one or more of the following herbs to custom blend an aromatic bath: comfrey, elderberry, horsetail, lavender, pine, rosemary and sage.

### Recipe:

—

2 cups fresh herbs or
1 cup dried herbs
2 gallons boiling water

*Steep herbs in water for 20 minutes.*
*Strain liquid and pour into a foot basin.*
*Soak feet for 20 minutes and pat dry.*

# Peppermint and Soy Oil Leg Rub

Peppermint, which has been found in tombs dating from 1000 BC, possesses anti-inflammatory, antispasmodic, stimulating, decongesting and soothing properties. Soy oil is an effective carrier oil and is absorbed quickly when applied to the skin. This rub provides relief for tired, congested and cramped legs.

## Recipe:

—

1/2 cup soy oil
4 drops peppermint essential oil

*Combine oils and massage
into tired legs. Follow with a
warm footbath. Pat dry.*

# Nighttime Hazelnut & Honey Hand and Foot Paste

This remedy is ideal for soothing hard-working hands and feet by gently exfoliating old, dead skin and nourishing the new cells. It will leave your hands and feet incredibly smooth.

## Recipe:

1/4 cup hazelnuts or almonds
1/4 cup dry oatmeal
3 tablespoons cocoa butter
2 tablespoons honey

*Process nuts in a blender or coffee grinder until coarsely ground. In a bowl, combine oatmeal, cocoa butter, honey and ground nuts. Rub into hands and feet and cover with cotton gloves and socks. Leave overnight. The next morning, remove gloves and socks and rinse hands and feet in cool water.*

# Apricot Oil Hand Moisturizer

Apricot oil is particularly effective on skin that
is prematurely aged, dry, sensitive or inflamed.

## Recipe:

6 dried apricots
2 cups water, preferably distilled
1 ounce grated beeswax
1 cup apricot oil
1/2 cup coconut oil
1/2 cup aloe vera gel
5 vitamin E capsules, broken open

*In a saucepan, bring water and dried apricots
to a boil and simmer for about 20 minutes.
Set aside to cool. In a double boiler set
over simmering water or in a heavy saucepan
set over low heat, combine the beeswax,
apricot oil and coconut oil and cook, stirring
constantly, for about 10 minutes. Pour into
a glass container and set aside. Place water
and apricots in a blender and blend until
smooth. Strain through a cheese cloth, discard
solids and put liquid back into blender.
Add aloe vera gel and vitamin E oil. Blend
at low speed. Slowly drizzle beeswax mixture
into blender and continue to blend at low
speed until mixture has the consistency of
frosting. You may not need to use all of
the beeswax mixture. Pour the mixture into
jars and store in a cool place. Mixture
will thicken as it cools. Apply to hands
after every washing.*

# CHAPTER 4

# Hair

**PRE-SHAMPOO TREATMENTS**

·

**SHAMPOOS AND CONDITIONERS**

·

**RINSES**

## *Four steps to beautiful hair:*

1. PRE-SHAMPOO TREATMENT: *to condition or moisturize hair, or to treat damaged hair.*

2. SHAMPOO: *to remove dirt, oil and residue.*

3. RINSE: *to remove any excess shampoo and to replenish the hair with nutrients.*

4. CONDITION: *to seal the moisture in the hair and protect it from harsh elements.*

# Beautiful and healthy hair comes from health within

- Like skin, hair needs protection from the sun: wear hats or protective sunscreen specially made for hair.

- The scalp is more responsive to treatments when the pores are dilated and when the skin is relaxed. A good time to apply a hair treatment is just before you get into a warm tub.

- On the final rinse, use cool water when rinsing your hair. It will tighten pores and leave your hair shinier.

- Use a broad-toothed comb instead of a hairbrush. The brush's bristles tend to damage the hair shaft, which can cause breakage.

- Do not overuse gels and mousses. They can cause build up and irritate the scalp.

- Beautiful and healthy hair comes from health within; if you eat well, drink plenty of water, get enough sleep and exercise and stay away from alcohol and cigarettes, it will show in your tresses.

*In the 1770's, French women would weave articles such as stuffed birds, jewelry, live flowers, vases and ship models into their hair pieces. Some of the wigs were so tall that doorways had to be raised to insure clearance.*

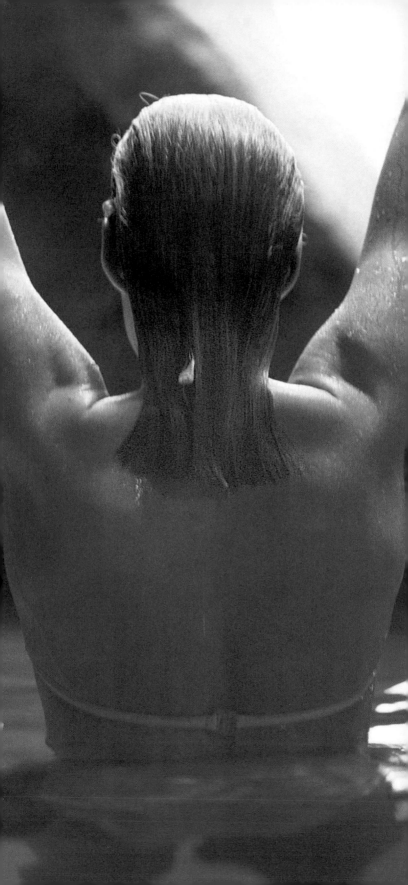

## Coconut Oil
## Hair Conditioning Clay

This hair treatment is both a deep cleanser and conditioner. The clay will remove dirt and hair product residue from the hair shaft while nourishing the scalp. Make sure you use the gentler white clay (such as kaolin), as other kinds of clay (red or green) might be too harsh for use on the hair. Coconut oil preserves the hair (and skin) by covering it with a thin layer that seals in natural moisture.

### Recipe:

—

1/2 cup jojoba oil
3 tablespoons coconut oil
4 tablespoons powdered white clay
1 cup distilled water

*In a small saucepan set over low heat, heat jojoba and coconut oils until melted. Remove from heat. Put powdered clay in a small bowl and slowly add water until it forms a paste. You might not need to use all of the water. Combine melted oils and clay and mix thoroughly. Apply a palm-full of clay treatment to clean, damp hair and massage vigorously into scalp. Cover hair with shower cap or with plastic wrap and leave on for 15 minutes. Shampoo hair as usual.*

•

## Hair grows about
## one-half inch per month.

# Mayonnaise & Avocado Hair Pack for Dry Hair

The combination of eggs and oil (from the mayonnaise) and avocado creates a wonderful conditioner for dry hair. If you have any extra time on your hands, it's always better to make your own mayonnaise. If you can't, then try to buy the all-natural product at a health food store.

## Recipe:

—

1 ripe avocado, peeled and pitted
1 cup mayonnaise

*In a small bowl, mash avocado and blend in mayonnaise. Gently massage into scalp and hair. Cover head with a shower cap or plastic wrap and leave treatment on for 20 minutes. Shampoo as usual.*

# Olive Oil & Lavender Conditioning Scalp Treatment for Normal Hair

Use this treatment once a week, especially if your hair has been over-exposed to sun and wind. Lavender oil has many healing and nourishing properties, and when combined with olive oil, which will improve the elasticity of the hair shaft and therefore make it less likely to break, this scalp treatment will leave your hair incredibly soft. The added bonus is that lavendar oil is known for alleviating headaches and tension when massaged into the scalp.

## Recipe

—

1/2 cup olive oil
10 drops essential lavendar oil

*Over low heat, warm olive oil in a small saucepan. Remove from heat and add lavender oil. Apply to damp hair while the oil is still warm and massagge into scalp. Cover head with a shower cap or plastic wrap and leave on for 20 minutes. Shampoo as usual.*

# Pre-Shampoo Wheat Germ Conditioner With Sage

Sage leaves are tonic and conditioning for hair. This treatment is particularly good for thinning and graying hair. It can also be used to help split ends by rubbing a bit of the treatment into the ends of the hair.

## Recipe:

—

8 sage leaves, crushed
1/2 cup boiling water
1 tablespoon wheat germ oil
1 tablespoon grapeseed oil

*Put crushed sage leaves in a small jar and cover with boiling water. Set aside for 30 minutes. Strain and reserve liquid and discard leaves. Add wheat germ and grapeseed oils to mixture. Pour into a dark-colored bottle. Rub a small quantity into your hair and scalp, preferably when hair is damp. Wrap with a towel and leave in for 45 minutes to an hour. Shampoo as usual.*

## Molasses
## Hair Wrap

This treatment is incredibly easy to do and will add shine and moisture to your hair. One word of caution: stay away from bees.

*Recipe:*

—

1/2 cup molasses (or honey or maple syrup)

*Apply molasses directly to damp hair and massage into scalp. Cover hair with shower cap or plastic wrap and leave in for 20 minutes. Rinse with warm water. Shampoo hair as usual.*

•

*Scientific studies show that hair grows faster in warm weather.*

# Daily Castile & Roman Chamomile Shampoo for Light Hair

As the name of this recipe suggests, this particular shampoo is very mild and can be used every day. While the use of essential oil in this mixture is optional, it will add a wonderful fragrance to your hair. You can also substitute fresh calendula or lemon peel for Roman chamomile and follow the same recipe.

## Recipe:

—

1/2 cup fresh (or 1/4 cup dried) Roman chamomile
1 1/2 cups boiling water
3 tablespoons pure Castile soap
1 teaspoon jojoba or sesame oil
3 drops patchouli or ylang-ylang essential oil
(optional)

*Crush herbs and put in a glass jar with a lid.
Pour boiling water over herbs. Let steep with
lid on for 20 minutes. Strain and reserve liquid
and discard herbs. Add Castile soap and jojoba
oil to chamomile infusion and stir well.
Add drops of essential oil and pour shampoo into
a flip-top or squeeze-type plastic bottle
for convenience.*

# Daily Castile & Rosemary Shampoo for Dark Hair

This recipe is particularly suited for dark hair. Sage or cloves can be substituted for rosemary.

## Recipe:
---

1/2 cup fresh rosemary (or 1/4 cup dried)
1 1/2 cups boiling water
3 tablespoons pure Castile soap
1 teaspoon jojoba or sesame oil
3 drops patchouli essential oil (optional)

*Crush herbs and put in a glass jar with a lid. Pour boiling water over herbs. Let steep with lid on for 20 minutes. Strain and reserve liquid and discard herbs. Add castile soap and jojoba oil to rosemary infusion and stir well. Add drops of essential oil and pour shampoo into a flip-top or squeeze-type plastic bottle for convenience.*

## Tea Tree Shampoo

This shampoo is particularly beneficial for individuals who have dry scalps. Tea tree oil has many healing and soothing properties, from treating burns and rashes to fighting viral infections. On the scalp, it will soothe dry, flaky skin.

*Recipe:*

—

2 tablespoons Daily Castile Shampoo
(see pages 139–140)
3 drops tea tree essential oil

*In a small bowl, mix shampoo and tea tree essential oil. Apply to hair and lather. Rinse and condition as usual.*

•

*The number of hairs a person has is determined by the number of hair follicles, which is established genetically before birth.*

*"Beauty is the power by which a woman charms a lover and terrifies a husband".*

—AMBROSE BIERCE

# Yogurt & Colorless Henna Conditioner for Brittle Hair

Colorless (neutral) henna is ideal for conditioning. It evenly coats the hair follicle with a fine film, sealing in the hair's natural oils and tightening cuticles to give hair more shine. Because the henna is colorless, it will not affect the color of your hair. The yogurt or sour cream adds an additional boost of conditioning. This treatment is not recommended if your hair is colored or chemically treated.

## Recipe:

1/4 cup colorless (neutral) henna powder
1/2 cup plain yogurt or sour cream

*Mix henna powder and yogurt in a small bowl. Apply to shampooed and damp hair, and cover head with a shower cap or plastic wrap and leave for 20 minutes. Rinse thoroughly and dry hair as usual.*

# Vitalizing Rosemary & Apple Cider Rinse

This after-shampoo rinse will remove all soapy residue from hair and leave it exceptionally shiny. Rosemary is often used in hair tonics and rinses as it stimulates the scalp and promotes hair growth.

## Recipe:

—

2 sprigs fresh rosemary
1 cup apple cider vinegar
3 cups warm water

*Chop rosemary and place it in a jar. Fill jar with apple cider vinegar and store in a cool dark place for 1 week, shaking jar vigorously once a day to disperse the rosemary. After 1 week, strain liquid, discard rosemary and return liquid to jar. When ready to use, dilute 1/3 cup of liquid in 3 cups of warm water and rub into freshly shampooed, damp hair and scalp. Avoid getting any mixture into your eyes. Rinse thoroughly and dry hair as usual.*

## Nourishing Hijiki Rinse

It is said that the wives of Japanese samurai used a hijiki-based shampoo to keep their hair strong and healthy. This rinse version is equally effective in that the seaweed not only removes residue but also nourishes the hair and scalp with a rich supply of nutrients.

### Recipe:

—

1 teaspoon powdered dried hijiki seaweed
1/2 cup distilled water

*In a small bowl, mix powdered seaweed and water until it forms a paste-like substance. Apply to clean, damp hair; massage into scalp and wrap hair with a towel. Leave on for 10 minutes, then rinse hair with cool water.*

"*Truth exists for the wise, beauty for the feeling heart.*"

—JOHANN VON SCHILLER

# CHAPTER 5

# Spa Cuisine

DRINKS

·

SOUP & SALADS

·

DESSERTS

"Broiling, roasting, steaming, boiling, poaching, grilling, low-fat, high-fiber, low-sodium, low-calorie, low-sugar, low-cholesterol, quality produce, fresh ingredients and, of course, delectable and delicious." These are the words generally associated with spa cuisine. As the cornucopia of cookbooks out there will attest, spa cuisine can even be—dare I say it?—mouth-watering and satisfying.

But one problem remains: if you're going to be soaking in seaweed, exfoliatingwith salts, or applying mud to your face all the while trying to relax, where oh where does one find the time to actually cook?

The recipes in this book will enable you to satisfy the craving for something healthy and good, with a minimal amount of prep time, cooking and cleaning.

So while you're waiting for that papaya face peel to take effect or that yogurt conditioner to work its magic, whip up a Blueberry Shiver or dive into a Beet and Haricots Verts Salad. Enjoy these luscious spa treats.

# Frozen Melon Bliss

SERVES ONE

### Recipe:

—

1/2 cantaloupe, diced and frozen
1/2 cup bottled or canned apricot nectar
1 teaspoon honey
1/2 cup chopped ice
banana slice for garnish

*Place cantaloupe, apricot nectar, honey and ice
in a blender and blend until smooth. Pour into
a glass and garnish with a slice of banana.*

# Orange Creamsicle Smoothie

SERVES ONE

### Recipe:

—

1 cup fresh orange juice
1/2 frozen sliced banana
1/2 cup nonfat vanilla frozen yogurt
1/2 cup chopped ice
slice of orange for garnish

*Place orange juice, banana, frozen yogurt
and ice in a blender and blend until
smooth. Pour into a glass and garnish
with a slice of orange.*

# Blueberry Shiver

SERVES ONE

### Recipe:

—

1/2 cup blueberries, frozen
1 banana, peeled and sliced
1/2 cup white grape juice
1/2 cup chopped ice
slice of banana for garnish

*Place blueberries, banana, grape juice and ice in a blender and blend until smooth. Pour into a glass and garnish with a slice of banana.*

# Honeydew-Lime Cooler

SERVES ONE

### Recipe:

—

1/2 honeydew melon, diced and frozen
juice of 2 limes
1 teaspoon honey
1/2 cup chopped ice
lime slice for garnish

*Place frozen honeydew, lime juice, honey and ice in a blender and blend until smooth. Pour into a glass and garnish with a slice of lime.*

# Citrus-Cranberry Sparkler

SERVES ONE

### Recipe:

—

1 32-ounce bottle sparkling mineral water
1 cup cranberry juice
1 orange, sliced
1 lemon, sliced
1 lime, sliced
1 cup ice cubes

*Empty mineral water into a large pitcher. Add cranberry juice and fruit slices. Serve over ice.*

# Kiwi Fizz

SERVES ONE

### Recipe:

—

2 medium kiwis, peeled
1 tablespoon honey
8 ounces mineral water, chilled
ice cubes
kiwi slice for garnish

*Place kiwis and honey in a blender and blend until smooth. Pour into a tall glass, add sparkling water and ice. Garnish with a kiwi slice.*

# Yellow Watermelon Gazpacho

SERVES TWO

### Recipe:

—

1 cup seeded, diced yellow watermelon
1 cucumber, peeled, seeded and finely chopped
1 red bell pepper, cored, seeded and
finely chopped
1 tomato, seeded and finely chopped
2 scallions, trimmed and finely chopped
1/2 red onion, finely chopped
1 rib celery, diced
1 tablespoon fresh lime juice
1 tablespoon balsamic vinegar
1/4 teaspoon salt
freshly ground pepper to taste
1 sprig parsley, trimmed and chopped, for garnish

*Put watermelon in a blender and pulse two
or three times until the watermelon is broken
up. Strain through a sieve, reserve juice and
discard pulp. In a bowl, toss cucumber, red
pepper, tomato, scallions, red onion, celery,
lime juice, vinegar, salt and pepper. Stir in
watermelon juice and chill for at least 1 hour
before serving. Garnish with parsley.*

# Warm Cucumber, Tomato & Lentil Salad with Balsamic Vinaigrette

SERVES TWO

### Recipe:

—

1/2 cup dry lentils
1 cucumber, peeled, seeded and diced
1 large tomato, seeded and diced
1/2 red onion, chopped
2 tablespoons fresh chopped parsley
3 tablespoons balsamic vinegar
2 tablespoons olive oil
1/4 teaspoon salt, freshly ground pepper to taste

*Cook dry lentils. When tender, drain and place in a large bowl. Add remaining ingredients. Toss well and serve warm.*

# Sweet Carrot & Coconut Health Salad

SERVES TWO

### Recipe:

—

2 cups grated carrots
1/2 cup sweetened shredded coconut
1/4 cup raisins
1/4 cup slivered almonds
2 tablespoons red wine vinegar
2 tablespoons olive oil
1/4 teaspoon salt
freshly ground pepper to taste

*Place ingredients in a large bowl and toss. Season to taste with pepper and serve.*

# Beet and Haricots Verts Salad

SERVES TWO

### Recipe:

—

2 large beets, boiled until tender, peeled and diced
1 cup haricots verts or very thin string beans,
lightly steamed
2 tablespoons chopped fresh parsley
1 shallot, finely chopped
1 1/2 tablespoons balsamic vinegar
2 tablespoons olive oil
1/4 teaspoon salt, freshly ground pepper to taste

*Place chopped beets and steamed haricots verts
in a bowl. Add parsley, shallot, balsamic
vinegar, olive oil, salt and pepper and toss.*

# Tricolor Salad with White Beans

SERVES TWO

### Recipe:

—

1 cup cooked and drained white beans
1 yellow pepper, stemmed, seeded and diced
6 cherry tomatoes, halved
2 tablespoons chopped fresh parsley
2 tablespoons red wine vinegar
3 tablespoons olive oil
1/2 teaspoon salt
white pepper to taste

*Place white beans, yellow pepper and cherry
tomatoes in a large bowl. Add parsley, vinegar,
olive oil, salt and white pepper and toss.*

# Red Bliss Potatoes With Fresh Dill Vinaigrette

### SERVES TWO

## Recipe:

—

6 small Red Bliss potatoes, scrubbed
2 tablespoons chopped fresh dill
2 1/2 tablespoons balsamic vinegar
3 tablespoons olive oil
1 teaspoon Dijon mustard
1/2 teaspoon salt, freshly ground pepper to taste
1 red onion, finely chopped

*Boil potatoes until tender, about 20 minutes.
Drain and quarter. In a blender, combine dill,
vinegar, oil, mustard and salt and blend. In a
large bowl, combine ingredients and dress.*

# Endive, Arugula & Pear Salad

### SERVES TWO

## Recipe:

—

1 1/2 teaspoons Dijon mustard
2 tablespoons balsamic vinegar
3 tablespoons olive oil
1/4 teaspoon salt, freshly ground pepper to taste
1 bunch arugula, rinsed and tough stems discarded
2 medium Belgian endive, trimmed and julienned
1 Bosc pear, peeled, cored and diced

*In a small bowl, blend mustard and vinegar.
Slowly whisk in olive oil, salt and pepper. Combine
arugula, endive and pear. Add dressing and toss.*

# Chilled Orange Slices in Honey

SERVES TWO

### Recipe:
—

2 seedless oranges, peeled and cut into wedges
(save some of the peel for garnish)
3 tablespoons honey
1 cup brewed orange pekoe tea, chilled

*Place orange wedges in a glass bowl.*
*Dissolve honey in the tea and add to bowl.*
*Chill for 4 hours before serving, garnished*
*with small strips of orange zest.*

# Cherry Vanilla Shake

SERVES ONE

### Recipe:
—

1 8-ounce can of pitted cherries, drained
1/2 cup nonfat plain yogurt
1 teaspoon honey
1/2 teaspoon vanilla extract
1/2 cup chopped ice

*Blend cherries in blender until completely*
*pureed. Add yogurt, honey, vanilla*
*extract and ice. Continue to blend until smooth.*
*Pour into a glass and serve.*

# Apricot Mousse

SERVES TWO

### Recipe:

—

1/2 cup dried apricots
2 tablespoons honey
1/2 cup low-fat or nonfat cottage cheese
1/2 cup nonfat plain yogurt
1/2 packet unflavored gelatin

*Put dried apricots in a small bowl and cover with boiling water. Set aside for 1 hour. Place drained apricots in a blender. Add honey and blend until smooth. In a small saucepan, dissolve gelatin in 2 tablespoons of water and place over low heat, stirring until the gelatin is dissolved. Combine cottage cheese and yogurt in a bowl and whip with a hand mixer or blend in a blender until smooth. Pour yogurt mixture into a bowl and slowly stir in the gelatin. Fold in apricot puree and pour into two glasses. Refrigerate for 1 hour before serving.*

# Papaya Boats with Lime-Honey Dressing

SERVES TWO

### Recipe:

—

2 tablespoons fresh lime juice
2 tablespoons honey
1/2 cup nonfat plain yogurt
1 ripe papaya, halved and seeded

*In a small bowl, combine lime juice and honey. Gradually stir in yogurt, making a creamy sauce. Pour into the center of the papaya halves.*

CHAPTER 6

# Essentials

OILS — ESSENTIAL & CARRIER

·

PAMPERING TOOLS

·

GLOSSARY

·

ASSOCIATIONS, RESOURCES AND SUPPLIERS

·

SPA LISTINGS

·

BIBLIOGRAPHY

·

CONVERSION TABLE

·

# Essential Oils

Essential oils are natural substances that are extracted from grasses, flowers, herbs, shrubs, trees, resins and spices, usually through a process called steam distillation. Oils can soothe, relax, rejuvenate, heal, energize or relieve pain, thereby affecting the body's physical, psychological and emotional levels. The use of oils in this manner is what is traditionally called aromatherapy.

## USES

• Essential oils can be inhaled through the nose and affect the brain via the olfactory system, or absorbed through the skin, sometimes reaching the bloodstream. Essential oils should never be applied directly to the skin and should always be diluted in a carrier oil such as jojoba, calendula or sweet almond (see below for suggested dilution rate) before being rubbed on the skin. (For a list of carrier oils, see page 176). Once diluted, they can be used for local applications on body parts or for full-body massages. They can also be added to bath water for a soothing experience or used in compresses to alleviate chronic pain (such as backache or arthritis) or acute pain (such as sprains or headaches). Essential oils can be mixed with water and alcohol for a refreshing room spray or vaporized in a burner to create an atmosphere or to eliminate unpleasant odors.

## INHALATION

• Place 5 drops on a tissue and inhale for 5 minutes; or add 6 to 8 drops to a bowl of almost boiling water, place a towel over your head and inhale for 5 minutes (excellent for colds/sinus infections as well as for facials); or put a few drops on your pillow before going to bed at night.

## MASSAGE TREATMENT

• Add 15 drops to one ounce of carrier oil and massage directly into the skin.

## SOOTHING BATH

• Add 5 to 10 drops to bath water and soak for 20 minutes; or add 10 to 15 drops to an ounce of coconut oil and pour into bath.

## SAFETY TIPS

• Essential oils should be stored in amber glass bottles away from direct sunlight and in a cool place. Never store an essential oil in a plastic bottle.

• If you store oils in the refrigerator, place the bottles in air-tight containers so that the aroma does not permeate food.

• Certain oils may solidify in cold temperatures due to their high wax content. If this occurs, place the oil bottle in a bowl of hot water to liquefy before use.

• Most essential oils have a shelf life of two years, with the exception of pine and citrus oils which lose some of their potency after 6 months.

• The color of certain oils may change with time; this does not affect the potency of the oil.

• Avoid using essential oils around the eye area.

• Never apply an essential oil directly to the skin; dilute it first (see below for dilution rates).

• Never use internally unless under the supervision and care of a specialist.

• Essential oils are not recommended for babies and small children and should always be stored out of the reach of children.

## COMPRESSES

• For chronic pain (aching joints, backache, sore and tired muscles, earache, etc.), prepare a hot compress by diluting 4 to 6 drops of oil in a large bowl of hot (almost boiling) water. Dip a cotton sheet strip about 4 inches wide into liquid and wrap around, or apply to, affected area. Repeat when compress has cooled off.

• For acute pain (sprains, headaches, etc.), prepare a cold compress by diluting 4 to 6 drops of oil in a large bowl filled with ice water. Dip a cotton sheet strip about 4 inches wide into the liquid and wrap around, or apply to, affected area.

## ROOM SPRAY

• Dilute 15 drops of oil into a half ounce of alcohol and add to a spray bottle filled with water. Use as an air freshener as often as desired.

## VAPORIZATION
• Add a few drops to a light bulb ring, fragrance burner or diffuser.

## DILUTION RATE
• The safest and most effective aromatherapy formulas are blended at a 2 1/2% concentration rate. Although the number of drops will vary slightly with the density of the oil and the diameter of the dropper opening, these measurements are sufficiently accurate for the purposes of the recipes in this book.

1. 2 to 3 drops of essential oil per teaspoon of carrier oil
2. 7 to 8 drops of essential oil per tablespoon of carrier oil
3. 15 drops of essential oil per ounce of carrier oil
4. 500 drops of essential oil per quart or liter of carrier oil

The following is a chart of the most essential of the essential oils and a brief description of their properties. Oils can be purchased at health food stores or by mail order. (See pages 190–191 for Shopping Guide.)

# Essential Oils

| NAME / LATIN NAME | PROPERTIES & SAFETY PRECAUTIONS |
|---|---|
| **Ajowan** *Trachyspermum copticum* | improves circulation, alleviates muscle pain<br>• *use sparingly on sensitive skin* |
| **Angelica** *Angelica archangelica* | strengthening, restorative, anchoring<br>• *avoid use in sun* |
| **Aniseed** *Pimpinella anisum* | aids in cramping, indigestion<br>or digestive problems<br>• *do not use if pregnant* |
| **Armoise** *Artimisia alba* | muscle relaxant, emollient<br>• *do not use if pregnant* |
| **Basil** *Ocimum basilicum* | soothing agent, muscle relaxant, toning<br>• *use sparingly* |
| **Bay** *Pimenta racemosa* | stimulating, energizing<br>• *can cause skin irritation* |
| **Bergamot** *Citrus bergamia* | skin conditioner, soothing agent, antiseptic<br>• *phototoxic* |
| **Birch Tar** *Betula lenta* | muscle relaxant, soothing agent<br>• *do not use if pregnant* |
| **Black Currant Seed** *Ribes nigrum* | relieves PMS, high source of vitamin C |
| **Black Pepper** *Piper nigrum* | muscle relaxant |
| **Cabreuva** *Myocarpus fastigiatus* | calming, increases alertness |
| **Cajeput** *Melaleuca cajuputi* | stimulating, mood improving, antiseptic |
| **Camphor** *Cinnamon camphor* | soothing agent, conditioner, muscle relaxant<br>• *do not use if pregnant or epileptic* |
| **Cananga** *Cananga odorata* | skin conditioner, deodorant |
| **Caraway** *Carum carvi* | muscle relaxant<br>• *slight dermal toxicity* |
| **Cardamom** *Elettaria cardamomum* | muscle relaxant, skin conditioner,<br>soothing agent |
| **Carrot Seed** *Daucus carota* | muscle relaxant, soothing agent,<br>skin conditioner |
| **Cedarwood Virginia**<br>*Juniperis virginiana* | antiseptic, skin conditioner, deodorant,<br>soothing agent |
| **Celery Seed** *Apium graveolens* | toner |
| **Chamomile Moroc** *Anthemis mixta* | muscle relaxant, skin conditioner |
| **Chamomile Roman** *Anthemis noblis* | muscle relaxant, skin conditioner |
| **Cinnamon Bark**<br>*Cinnamomum zeylanicum* | skin conditioner, anti-inflammatory agent<br>• *can cause skin irritation* |
| **Citronella** *Cymbopogon nardus* | skin conditioner, insect repellent |
| **Clary Sage** *Salvia sclarea* | skin conditioner, astringent,<br>soothing agent, muscle relaxant<br>• *do not use if pregnant; do not drink alcohol<br>or drive* |
| **Clove Bud** *Syzgium aromaticum* | muscle relaxant, soothing agent<br>• *can cause skin irritation* |
| **Copaiba Balsam** *Copaifera officinalis* | increases circulation, reduces stress |

| NAME / LATIN NAME | PROPERTIES & SAFETY PRECAUTIONS |
|---|---|
| **Coriander** *Corriandrum sativum* | muscle relaxant, soothing agent<br>• *use sparingly* |
| **Costus Root** *Sassuriea costus* | calming |
| **Cumin** *Cuminum cyminum* | stimulating<br>• *can cause skin irritation* |
| **Cypress** *Cupressus sempervirens* | antiseptic, astringent, soothing agent,<br>skin conditioner<br>• *flammable* |
| **Cypriol** *Cyperus scariosus* | aids digestion |
| **Eucalyptus** *Eucalyptus globulus* insect repellent | antiseptic, soothing agent, skin conditioner, |
| **Evening Primrose** *Centhera biennis* | good for dry skin and eczema |
| **Fennel** (sweet)<br>*Foeniculum vulgare dulce* | muscle relaxant, soothing agent, antiseptic<br>• *use sparingly* |
| **Frankincense** *Boswellia carteri* | skin conditioner, soothing agent |
| **Galbanum** *Ferula galbaniflua* | skin conditioner, muscle relaxant |
| **Geranium** *Pelargonium graveolen* | skin refresher, astringent |
| **Ginger** *Zingiber officinale* | astringent |
| **Grapefruit** *Citrus paradisi* | soothing agent, astringent, skin conditioner |
| **Hyssop** *Hyssopus officinalis* | soothing agent, skin conditioner<br>• *do not use when pregnant, if suffering from epilepsy or high blood pressure* |
| **Jasmine Absolute**<br>*Jasminum officinale* | emollient, soothing agent, antiseptic |
| **Juniper** *Juniperus communis* | skin detoxifier, astringent, soothing agent<br>• *flammable* |
| **Labdanum** *Cistus ladanifer* | skin conditioner |
| **Lavandin** *Lavandula hybrida* | soothing agent, muscle relaxant,<br>skin conditioner, astringent |
| **Lavender** *Lavandula officinalis* | muscle relaxant, skin conditioner,<br>soothing agent, astringent |
| **Lemon** *Citrus limonum* | soothing agent, antiseptic |
| **Lemongrass**<br>*Cymbopogon flexuosus* | skin conditioner, soothing agent,<br>muscle relaxant, antiseptic<br>• *can cause skin irritation* |
| **Lime** *Citrus aurantifolia* | soothing agent, skin conditioner, astringent |
| **Mandarin** *Citrus reticulata* | soothing agent, astringent, skin conditioner |
| **Manuka** *Leptospermum* | relieves aches and pains,<br>healing to the skin |
| **Marjoram** *Origanum marjorana* | antiseptic, calming |
| **Mimosa** *Acacia dealbata* | muscle relaxant, skin conditioner,<br>soothing agent |
| **Myrrh** *Commiphora myrrha* | anti-inflammatory agent, emollient, antiseptic<br>• *use in moderation if pregnant* |

| NAME / LATIN NAME | PROPERTIES & SAFETY PRECAUTIONS |
|---|---|
| **Myrtle** *Myrtus communis* | soothing agent, astringent, skin conditioner, muscle relaxant |
| **Neroli** *Citrus aurantium* | antiseptic, emollient |
| **Nutmeg Myristica fragrans** *Niaouli elaleuca viridiflora* | antiseptic, soothes irritated skin, muscle relaxant<br>• **use sparingly** |
| **Orange** *Citrus sinensis* | astringent, soothing agent, skin conditioner |
| **Origanum** *Origanum vulgare* | increases energy<br>• **can cause skin irritation** |
| **Palmarosa** *Cymbopogon martini* | skin conditioner, soothing agent, emollient, muscle relaxant |
| **Patchouli** *Pogostemon cablin* | anti-inflammatory agent, antiseptic, astringent |
| **Peppermint** *Mentha arvensis* | emollient, antiseptic, muscle relaxant<br>can cause skin irritation |
| **Petitgrain** *Petitgrain bigarade* | relieves anxiety and stress |
| **Pine** *Pinus sylvestris* | antiseptic<br>• **can cause skin irritation** |
| **Rose Absolute** *Rosa damascena* | skin conditioner |
| **Rose Otto** *Rosa —* | astringent |
| **Rosemary** *Rosmarinus officinalis* | antiseptic, muscle relaxant, soothing agent, skin conditioner<br>• **do not take if pregnant or have high blood pressure** |
| **Rosewood** *Aniba rosaeodora* | muscle relaxant |
| **Sage** *Dalmatian Salvia officinalis* | soothing agent<br>• **do not use if pregnant or suffering from epilepsy** |
| **Sandalwood** *(Mysore) Sandalum album* | antiseptic, emollient, soothing agent, astringent, skin conditioner |
| **Spearmint** *Mentha spicata* | emollient, astringent, soothing agent, muscle relaxant<br>• **use sparingly** |
| **Tarragon** *Artimisia dracunculus* | astringent |
| **Tea Tree** *Melaleuca alternifolia* | antiseptic<br>• **may cause irritation to sensitive skin** |
| **Thyme** *Thymus vulgaris* | antiseptic, toner<br>• **can cause skin irritation** |
| **Vanilla** *Vanilla planifolia* | emollient |
| **Vetiver** *Vetiveria zizanioides* | emollient, reduces blood pressure |
| **Violet Leaves** *Viola* | soothing agent, skin conditioner |
| **Yarrow** *Achillea millefolium* | reduces scarring |
| **Ylang-Ylang** *Cananga odorata* | reduces stress and tension |
| **Zanthoxylum** *Zanthoxylum alatum* | reduces stress and tension |

# Carrier Oils

Because essential oils cannot be applied directly to the skin in their pure state, they must be diluted in carrier oils, also known as "base oils." These are rich in fatty acids, vitamins and moisturizing nutrients. In addition to providing good lubrication for the massage therapist's hands against the patient's skin, carrier oils contain important healing substances that render the skin more elastic and soft. They can be added directly to bath water for a skin-nourishing soak, used in hair treatments and in numerous skin applications. (See page 171–174 for a list of essential oils, their properties, and for dilution rates.)

For a personalized massage mixture, add 15 drops of an essential oil per ounce of carrier oil. Once blended, store the mixture in a glass bottle in a cool, dark place. These oils will generally keep for 12 months.

The following is a list of the most common carrier oils. They can be purchased in health food stores or through mail order. (See pages 190–191 for Shopping Guide.)

# Carrier Oils

| NAME | BOTANICAL NAME | PROPERTIES |
|------|----------------|------------|
| **Almond** (sweet) | *Prunus amygdalus* | for all skin types; softens skin, relieves itchiness |
| **Apricot kernel** | *Prunus armeniaca* | excellent for facial massages; rehydrates, restores skin's glow |
| **Avocado** | *Persea americana* | for dry skin; easily absorbed, plumps up prematurely lined skin |
| **Borage** | *Borage officinalis* | rich in gamma linolenic acid; treats eczema and psoriasis; anti-aging properties |
| **Calendula** | *Calendula officinalis* | heals cracked skin and rashes |
| **Carrot Seed** relieves | *Daucus carota* | rejuvenates; reduces premature aging and scarring, itching, restores elasticity |
| **Coconut** | *Cocos nucifera* | excellent for cracked or brittle skin; helps retain skin's moisture |
| **Evening Primrose** | *Oenothera fiennis* | for dry skin; high in essential fatty acids |
| **Grapeseed** | *Viti vinifera* | good for full-body massage; odorless and easily absorbed |
| **Hazelnut** | *Corylus avellana* | good for oily skin; easily absorbed |
| **Jojoba** | *Simmondsia chinensis* | anti-bacterial, anti-inflammatory, hydrating, good emollient |
| **Macadamia** | *Macadamia integrifolia* | helps maintain the natural moisture level of the skin |
| **Olive Oil** | *Olea europaea* | disinfecting, healing, excellent to soothe chapped skin |
| **Passion fruit** | *Passiflora incarnata* | helps maintain skin's elasticity |
| **Rosehip** | *Rosa rubiginosa* | promotes tissue regeneration; good for scars, burns and wrinkles |
| **Safflower** | *Carthamus tinctorius* | light texture; easily absorbed |
| **Sesame** | *Sesamum indicum* | good natural sunscreen; practically odorless |
| **Sunflower** | *Helianthus annuus* | softens skin; helps skin maintain its natural level of moisture |
| **Wheat Germ** | *Triticum vulgare* | aids muscle and lymph function; good for dry skin; natural antioxidant, strong odor |

# Pampering Tools

The ultimate home spa experience requires two sets of Pampering Tools.

The first set, "The Basics," can be found in most kitchen cupboards. These are the whisks, the empty jars, mixing bowls and such for preparing and storing masks, moisturizers, rinses and other treatments.

The second set of tools, "Mood Enhancers," will provide the added personal touches that will bring your home spa experience to the highest dimension of complete relaxation, rejuvenation and luxury: from the oversized, ultra-soft terrycloth robe to the inflatable neck pillow for reading while soaking in the tub.

*Note:* Try to stay away from any tools that have silver, copper, aluminum, Teflon or cast-iron finishes, as some ingredients you will be using react with these materials. Stainless steel, enamel or glass are preferable.

# Pampering Tools

## THE BASICS

**blender** or **food processor**
*for preparing lotions, moisturizers, cleansers, and grinding herbs. A glass blender is easier to clean than a plastic one.*

**spice mill** or **coffee grinder**
*for finely grinding herbs.*

**wire whisk** *for blending dry ingredients (bath salts or herbs) and for evenly distributing oils and scents in creams and lotions.*

**tongs** *especially useful for picking up hot sheet strips used for body wraps.*

**double boiler** *for evenly heating and melting without scorching.*

**lobster pot** or **boiling pot** *to prepare body wraps or to heat towels and sheets.*

**glass bowls** *for mixing ingredients.*

**spa bowls** *2- or 3-quart plastic or ceramic bowls to hold seaweed, clay, salts and/or water.*

**large bowl** or **basin** *for facial steams.*

**wooden spoons** *to mix, stir and blend.*

**stirring rod** *for stirring while heating; glass is preferable but a chopstick works too.*

**vegetable peeler** *for peeling oranges and lemons.*

**knives** and **chopping board** *for cutting and chopping.*

**funnel** *to facilitate bottling lotions, moisturizers, rinses, etc.*

**glass (or plastic) jars** and **bottles** *for storing your preparations (moisturizers, creams, lotions, masks, and bath salts). Pickle, mustard, mayonnaise or jam jars are perfect for body lotions and creams. Smaller, wide-mouthed jars are better for face moisturizers or masks.*

**strainer** *for straining herbs from solutions and mixtures.*

**plastic bottles** *flip-top or squeeze-type bottles are best for storing conditioners and shampoos. Spray bottles are good for rinses.*

**glass mason** or **canning jars** *for brewing teas and making infusions.*

**small glass amber bottles** *for storing essential oils.*

**measuring cup** *preferably glass because it's easier to read and clean.*

**measuring spoons** *standard set.*

**dropper** *preferably glass, for accurately measuring essential oils.*

**loofah sponge** or **mitt** *natural or synthetic fiber, for applying body treatments.*

**natural bristle brush** *a short-handled brush with a head about the size of your palm for skin treatments.*

**pumice stone** *for exfoliating and foot scrubbing.*

**plastic wrap, shower cap** or **plastic bag** *for hair treatments that need to be left on for an extended period of time. The plastic intensifies the effects of the heat that is useful for hot oil hair treatments and for certain conditioners.*

**disposable rubber gloves** *for use when handling mud or seaweed for wraps or masks, or when applying hair treatments.*

**towels** *two or three oversize bath towels to cover or wrap your body with, two standard-size towels to use in steam facials or to wrap your hair in, two hand towels to pat dry with and two washcloths to be used as compresses.*

## THE BASICS

**two (or more) cotton bed sheets** *for lying on and covering yourself with to maintain body heat after a massage or wrap; unbleached, all-natural fiber is preferable.*

**two cotton sheets** *one to cut into wrapping strips for body wraps, another (a twin flat sheet is a good size) for the full body wrap; unbleached, all-natural cotton is preferable.*

**pillows** *to raise your feet or your head; goose down is usually more comfortable, but synthetic fibers are easier to machine wash in case of spills.*

**blanket(s)** *for staying warm after a wrap, massage or bath, and to be used over the cotton sheet. Although wool is certainly warmer, it's best to get a cotton blanket that can be easily washed if you spill lotions or oils on it.*

**shower curtain** or **plastic sheet** *to retain the heat of a body wrap.*

**muslin sacks** *for preparing herbal bath sachets, soaks and teas.*

## MOOD ENHANCERS

**candles** *chose an aromatherapy candle to enhance the effect of the bath and accentuate the mood.*

- *balancing Lavender or Geranium*
- *meditation Cedarwood, Pine or Patchouli*
- *healing Rosemary, Eucalyptus or Cedarwood*
- *stress relief Lavender, Grapefruit or Bergamot*
- *headache relief Peppermint, Lavender, Marjoram*

**music** *soothe your ears while soothing your body.*

**books** *soaking is the time to fully indulge with whatever tickles your reading fancy: thrillers, romance novels, classics of the stage or great novellas.*

**inflatable neck pillow** and **bathtub tray** *to add comfort while soaking.*

**space heater** *keeps the body warm after a massage or body wrap.*

**WARNING:** THIS INSTRUMENT SHOULD ONLY BE USED UNDER DIRECT SUPERVISION AT ALL TIMES. NEVER LEAVE THE SPACE HEATER UNATTENDED.

**beverage of choice** *a pitcher of mineral water with lemon, lime and orange slices; herbal tea, or health drink (see pages 149–152 for recipes). Stay away from alcohol as it will increase body temperature and heart rate.*

**slippers** and **your favorite bathrobe** *for comfort and warmth after treatments.*

**fresh cut flowers** *for tubside*

**answering machine** *turn it on, put the volume on low and turn off the phone ringer.*

# Pampering Tools

# GLOSSARY OF TERMS AND TECHNIQUES

**acupressure** An ancient Chinese technique based on the theory that there are meridians (pressure points) on the body that correspond to different organs in the body. When pressure is exerted on a particular meridian, energy pathways are stimulated for that particular organ to which the meridian corresponds and contracted muscles are encouraged to relax.

**acupuncture** An ancient Asian healing technique which targets energy meridians through the use of fine needles or low-voltage electric current (electro-acupuncture). Its purpose is to relieve arthritic, neurological and muscular tensions.

**adaptogen** A substance that stimulates the body's immune mechanism and helps it adapt to a new stress.

**Alexander Technique** A bodywork system created in the 1890s by an Australian by the name of F.M. Alexander. Its purpose is to help improve posture and to correct bad physical habits that cause stress on the body.

**alterative** Restores the normal functions of a body organ or system; promotes healing.

**analgesic** Pain reliever.

**antifungal** A substance that clears and counters fungal infections.

**antioxidant** A substance that prevents other substances (like fats) from binding with oxygen, thereby slowing their deterioration process.

**aromatherapy** The art of using aromatic essences (essential oils) derived from flowers, leaves, roots, woods and fruits. These essences can be used for various treatments including total body massage, inhalation therapy and baths. During the time of the plague in Europe, people who worked closely with essential oils (pharmacists or priests, for example) seemed to be spared from contracting the disease. The word "aromatherapy" was first coined by the French in the 1930s.

**astringent** Constricts tissue to cut off the flow of fluids or blood, thus giving skin a smoother and firmer texture and appearance.

**Ayurveda** An ancient Indian system of traditional folk medicine that incorporates nutrition, essential oils, massage and meditation to restore the body to its perfect balance.

**Bach Cures** The art of healing with floral essences and oils.

**Balneotherapy** Water-based treatments (such as those using thermal hot springs, mineral water or seaweed which helps digestion and relieves gastric discomfort.

**carrier oils** Oils such as jojoba oil, sweet almond oil, or apricot kernel oil used as a base in which aromatherapy oils are mixed. Aromatherapy oils are so concentrated that they should rarely be used more than a few drops at a time and should be diluted in a base.

**Cathiodermie** A rejuvenating treatment for the skin using electric stimulation in minimal doses.

**Chi Kung** A Chinese exercise focusing on breathing and body movements to recharge and energize the body.

**cold plunge** A deep pool of water in which the water temperature is kept to 60 degrees F (15.5 degrees C) or below, usually used after a steamy hot shower. A quick plunge in the water will invigorate instantly.

**colonic irrigation** A water enema that cleanses high into the colon.

**compress** A cloth soaked in a hot herbal infusion which, when applied to the affected area, relieves irritation and fatigue. Compresses may also be cold.

**cranial massage** A soothing manipulation of the pressure points of the spine, shoulders, neck, and head. Its purpose is to improve the flow of cerebral spinal fluid throughout the nervous system to relieve headaches, back problems and teeth grinding.

**crystal healing** A healing energy generated by quartz and other minerals.

**Dead Sea mud treatment** A mineral-rich mud imported from the Dead Sea applied to the body to cleanse pores and relax muscles. The mud, especially high in magnesium chloride, potassium chloride and calcium, will ease arthritic and rheumatic pain.

**decoction** A solution made by simmering roots, barks, or woody parts of plants in water for a period of time to extract active ingredients and then straining the infusion.

**demulcent** Substance that soothes and heals mucous membranes (inside of mouth, throat, nose).

**detoxification** A process that speeds up or facilitates the elimination of toxins from the body and increases circulation. Often heat, water and herbs are used. Detoxification therapies include steam, sauna, inhalation therapy, and whirlpool baths.

**diaphoretic** A substance that induces perspiration.

**diuretic** A substance that increases the release and elimination of toxins, generally through urination.

**draping** The use of sheets or towels to cover a patient during a massage.

**dulse scrub** To remove dead skin and to enrich the skin with vitamins and minerals, the body is scrubbed with a mixture of powdered dulse seaweed and oil or water.

**effleurage** Massage term used to describe a long stroking motion intended to calm the nervous system

**electro-acupuncture** Acupuncture-type treatment that utilizes low-voltage electric current instead of fine needles.

**electrotherapy treatments** that use ultrasound, short waves, infrared rays or various forms of electricity for results.

**emollient** A substance which soothes and softens.

**energy balancing** A technique that involves clearing and charging of the seven major chakras and auric levels. A powerful current of healing energy comes through the therapist's hands to heal areas of injury and illness.

**Esalen Massage** A modern variation of the Swedish massage, developed at the Esalen Institute in Big Sur, California. Its focus is not so much on relieving muscle tension or increasing circulation as it is on creating deeper states of relaxation, beneficial states of consciousness and general well-being. Whereas Swedish is brisk and focuses on the body, Esalen is slower, rhythmic and hypnotic, and focuses on the mind/body as a whole.

**essential oils**  Oils, such as peppermint, sandalwood or tea tree, that are used in aromatherapy. (See section on essential oils, pages 168–178)

**European facial**  A massage technique that cleanses and steams the face, shoulders and chest area. Its purpose is to nourish and refine skin.

**exfoliation**  A treatment whose primary purpose is to scour dead skin cells from the body; comes from the Latin word *exfoliare*, which means "to remove leaves."

**facial**  A deep-cleansing treatment of the face, neck and shoulders.

**fango therapy**  A combination of hot paraffin wax and natural volcanic ash spread over the entire body (or parts of it) to promote joint and muscle mobility and relaxation and alleviate aches and pains due to illness or injury. *Fango* is Italian for "mud."

**Feldenkrais Method**  Created by Moshe Feldenkrais, a Russian-born Israeli educator, this "movement therapy" uses gentle manipulation of the muscles to help reprogram them to work more efficiently. Subtle changes are introduced to the person through touch, which will help to break up old patterning.

**flotation tank**  A tank filled with mineral-enriched water. Floating in darkness in this tank promotes relaxation.

**foot (or hand) reflexology**  A technique that maintains that the body is divided into 10 zones, all of which have a corresponding reflex area on the foot (or hand). Applying pressure to a particular massage point on the foot (or hand) helps circulation, promotes relaxation and relieves pain in one of the 10 body zones.

**gommage**  A cleansing and rehydrating treatment through the use of creams which are applied in long, massage-type movements.

**guided imagery**  Visualization to stimulate the body's immune system.

**Hakomi Method**  A body-based psychotherapy using special states of consciousness to help probe non-verbal levels. Body-mind awareness and touch are used to explore the body as a deep source of information.

**haysack wrap**  Body detoxification treatment through the use of steamed hay.

**Hellerwork**  A system of deep-tissue body work, stress reduction and movement re-education developed by Joseph Heller.

**herbal wrap**  The body is wrapped in sheets soaked in a heated herbal solution, then covered in plastic blankets or sheets. The heat eliminates impurities from the skin and enhances muscle relaxation.

**holistic health**  A philosophy which seeks to achieve balance and harmony and to promote well-being by attending to elements of life such as emotional, spiritual and physical health as well as lifestyle.

**homeopathy**  Form of medicine in which patients are treated with small quantities of natural substances which trigger symptoms much like those that are in need of being cured. The body is thus encouraged to heal itself.

**humectant**  A substance that promotes the retention of moisture.

**Hydroculator**  A pre-made clay-filled compress.

# Glossary

**hydrotherapy** Water treatments (underwater massages, hot and cold showers, mineral baths and jet sprays, for example) which assist in the healing process of almost any ailment. This therapy, which is one of the oldest methods known to man, provides both stimulation and relaxation at the same time.

**hypnotherapy** Engendering a state of physical and mental relaxation which is applied to psychological issues.

**infused oil** An aromatic infusion of herbs in oil applied externally or used in creams or salves.

**inhalation room** A room in which steam is mixed with eucalyptus. Inhalation of the steam is meant to decongest the respiratory system.

**iodine-brine therapy** A treatment that uses bathing in water rich in iodine and salt as a method of facilitating the recuperation and convalescence process.

**iridology** A method of detecting the condition of body organs by reading markings in the iris of the eye.

**Jin Shin Jyutsu** A gentle Japanese massage technique which is based on synchronizing the pulses in two parts of the body, or two meridians, thus releasing natural pain relievers and rejuvenating the immune system.

**Kneipp System** Named after Father Sebastien Kneipp, these treatments combine hydrotherapy with herbal preparations and a diet of natural foods.

**Laciol manicure** Hands are soaked in a dish of milk or in a warm moisturizing cream to soothe and smooth skin.

**Lomi-Lomi** A Hawaiian rhythmical rocking massage.

**loofah scrub** A full body scrub with a loofah sponge and sea salt usually mixed with a warm oil (avocado or almond). Its purpose is to exfoliate the skin and to stimulate circulation.

**lymph drainage** A gentle and pulsating pressure or massage of specific areas of the body located around the lymph nodes and toward the heart. The purpose of this massage—used extensively for neck, head, and shoulders during facial massage—is to increase the lymphatic flow to promote the body's internal cleansing (toxin drainage).

**massage** Muscle manipulation (that includes acupressure, polarity, reflexology, effleurage, stroking, kneading, friction or rocking) intended to stimulate circulation, increase suppleness and reduce stress.

**moor mud pack** A healing treatment prepared with nutrient-rich muds from the bottom of inland spa lakes. Its purpose is to improve cell regeneration, stimulate lymph glands and balance hormones.

**mud bath** The body is coated with organic thermal mud to release tension and nourish the skin with minerals.

**Myofacial Release** A gentle release, through soft tissue manipulation, of the body's facial system to restore proper function and bio-mechanics.

**naturopathy** Natural healing prescriptions that use plants and flowers.

**neuromuscular therapy** A technique that concerns itself with bringing relief from soft tissue pain and dysfunction. The focus of the work is to find the origin of the pain and address it. NMT renews structural homeostasis by restoring normal physiological functioning among muscles, nerves and the musculoskeletal system, and is very effective for breaking chronic pain cycles.

**Ortho-Bionomy Method**  A technique developed in the 1970s by the body worker Arthur Lincoln Pauls. This approach uses gentle, relaxing movements and postures to help the body release tensions and muscular holding patterns, without use of force or pressure. Its goal is a restoration of structural alignment and balance.

**Panchakarma**  A system or way of life which focuses on meditation, diet and cleansing and purifying treatments to achieve an internal balance as well as a balance with one's surroundings.

**Parafango**  A combination of mud and paraffin wax. (See fango therapy entry above)

**paraffin treatments**  Hands, feet or the entire body are dipped in warm paraffin wax, a complex mixture of hydrocarbons and then covered in plastic wrap. Wax is peeled off after it hardens, about 15 minutes later. Besides leaving the skin smooth, ingredients in the wax moisturize and increase joint and muscle mobility.

**peliotherapy**  The therapeutic use of muds.

**peloids**  A generic term for muds.

**percussion**  A massage term for gentle and rhythmical taps on the body, akin to a light karate chop, the purpose of which is to awaken the body and increase its vitality.

**petrisssage**  Massage term for a deep circular movement of the fingertips or thumbs on a particular muscle.

**Pfrimmer Therapy**  Developed by Therese Pfrimmer, this technique works across the muscles, manipulating deep tissues, and is aimed at stimulating circulation and regenerating lymphatic flow in order to promote detoxification and oxygenation of stagnant tissues.

**phytotherapy**  A generic term used for treatments through mud packs, baths, massages, or inhalation using natural herbs, plant oils or extracts.

**Pilates Method**  A body conditioning system which focuses on improving flexibility and overall body strength without building bulk. Developed in Germany by Dr. Joseph Pilates during the 1920s.

**Polarity Therapy**  Founded by osteopath Dr. Randolph Stone, a massage technique combining deep pressure massage with gentle rocking and stretching. The purpose of the method is to achieve correct body alignment.

**poultice**  Plant or herb compress aimed at relieving pain, swelling or irritation. The plant or herb itself is applied directly to the affected area, then a cloth is wrapped over and around the plant or herb.

**reflexology**  A system of massaging specific to the hand or the foot that promotes healing, improves circulation, and relieves stress in other parts of the body.

**regenerative**  Reviving or producing tissue growth.

**Reiki**  This is an energetic approach to the body, where the therapist places his or her hands on, or just above, 12 proscribed areas of the body. In each position the hands are rested gently on the body and remain still for

3 to 5 minutes without manipulation of any sort. The purpose of this therapy is to promote deep relaxation by creating a calming and nurturing effect.

**repechage** Facial or full body cleansing and moisturizing treatment using clay or mud, herbs and/or seaweed.

**rocking** A massage technique in which the hands gently rock or shake the body back and forth.

**Rolfing** Created by Ida Rolf, this intensive, deep and sometimes painful massage technique is used to realign the skeletal structure, improve energy flow and relieve stress caused by emotional trauma.

**Roman Bath** A heated seawater Jacuzzi with jets and benches for seated bathing.

**Roman Chair** Exercise machine specifically intended for strengthening the back.

**Rosen Method** A technique developed by Marion Rosen that uses gentle, non-intrusive touch and verbal exchange between practitioner and client to help draw the client's attention to areas of stress or tension. This serves to help the client become fully aware of how the patterns of tension are associated with emotional or unconscious material. This awareness itself is the key that allows the tension or holding patterns to be released.

**rubefacient** Causing redness of the skin.

**Russian Massage** Massage technique that requires precise angles for the muscles and joints. The deep-tissue massage aims to stretch the muscles out.

**salt glow scrub** A mixture of coarse salt and aromatic oils rubbed all over the body to help shed flaky and drying skin and improve skin circulation and texture.

**salve** An herbal ointment that is not water-based and does not blend with the skin but forms a layer over it. It is used to protect, nourish or treat sensitive or injured skin.

**sauna** Dry heat used to open the pores and encourage the sweating out of impurities.

**Scotch Hose** A high-powered water spray using hot or cold seawater or freshwater. An excellent invigorating, energizing and cleansing treatment.

**seaweed face mask** or **seaweed wrap** The face or body is covered in mineral-rich seaweed which is then rinsed off, leaving the skin smooth, revitalized, remineralized and restored.

**Shiatsu** A Japanese massage technique in which pressure is gently applied along the meridian points of the nervous system's energy paths. The purpose of this massage is to revitalize and balance the endocrine and immune systems and to stabilize energy flow.

**sports massage** A deep tissue massage —often around the joints— for treating specific muscle groups.

**stress management** A program of meditation and deep relaxation intended to reduce the effects of stress on the system.

**structural integration** A system that relieves the patterns of stress and impaired body functioning due to poor posture, chronic and acute conditions (such as lower back pain, or neck and shoulder injuries) through the manipulation of deep and superficial connection tissue, plus movement education.

**sweat lodge** A Native American spiritual purification ceremony.

**Swedish Massage** This European massage technique was developed in the late 18th century by a Swedish fencing master. It combines various techniques including stroking, kneading, manipulation and tapping of muscle tissue. One of the primary goals of Swedish massage is to speed the venous return of unoxygenated and toxic blood from the extremities and to shorten recovery time from muscular strain by flushing the tissues of lactic acid, uric acid and other metabolic wastes. It increases circulation without increasing heart load in order to relax muscles and improve overall circulation.

**Swiss shower** Powerful water jets aimed at different parts of the body to create an invigorating massage.

**T'ai Chi Ch'uan** An ancient Chinese martial art form of stylized, meditative exercise characterized by methodically slow circular and stretching movements and positions of bodily balance.

**tapotement** A massage technique that involves a light, steady tapping that causes a slight vibration of the muscle.

**thalassotherapy** Water-based treatments using seawater, seaweed, algae and/or mud to rejuvenate, energize, cleanse, nourish and detoxify) which improve circulation, detoxify and revitalize the body.

**tincture** An infusion in which a particular herb or blend of herbs is steeped in alcohol, glycerin, or vinegar and water.

**tonic** A substance that fortifies the body when taken over a long period of time and balances the "qi," what the Chinese call "the vital life force."

**Tragerwork** Developed by Milton Trager over 65 years ago, this meditative method entails rhythmic and repetitive movements which trigger tissue change.

**Vichy shower** An energizing treatment with water jets of varying temperatures and pressures.

**Visceral Manipulation** A form of manipulation that focuses on the internal organs' environment as well the organs' influence on many structural and physiological dysfunctions.

**Vodder Massage** A manual lymph drainage technique developed by Danish-born Emile Vodder in the 1950s.

**Watsu** Underwater Shiatsu massage.

**ying/yang theory** An ancient Chinese philosophy based on two distinct presences which are apparent in every aspect of life: Yin (the dark side of the mountain) and Yang (the sunny side). On the body, the Yang meridians are on the side that "faces" the sun (the back) and the Yin meridians are on the "shadow" side (the front of the body).

**yoga** A discipline of stretching and toning the body through movements, breathing exercises, postures, relaxation techniques, and diet. A technique practiced to achieve control of the body and mind.

**Zen Shiatsu** A Japanese art that uses finger pressure (acupressure) to unblock and release "energy channels," producing general well-being.

**Zero Balancing** Created by Fritz Smith, MD, this is a system that focuses on guiding the body to find its own equilibrium by balancing body energy and body structure.

## BASIC INFORMATION: ASSOCIATIONS, RESOURCES AND SUPPLIERS

**MASSAGE THERAPIES**

About Massage Magazine
www.massagemag.com

Acupuncture.com–Your Online Resource for
Traditional Chinese Medicine
www. acupuncture.com

Traditional Chinese Medicine
www. acupuncture.com

Alive! Therapeutic Massage & Bodywork:
Reiki Healing Center
www.massagetherapy.com

Alternative Health Care Definitions
www.althealthsearch.com

American Massage Therapy Association
www.amtamassage.org

Ayurvedic Resource List
www.ayur.com

Canadian Massage Therapist Alliance
www.collinscam.com

HealthWorld Online: Varieties and Techniques
www.healthy.net

Home of Reflexology
www.reflexology.org

International Massage Association
www.imagroup.com

Massage Association of Australia
www.maa.org.au

Massage for Health
www.assaetherapy.net

Massage Network: Different Types of Massage
www.massagenetwork.com

Massage Therapists' Association
of British Columbia
www.massagetherapy.bc.ca

Massage Therapy Network

www.massagetherapy.net
Massage Therapy Web Central
www.qwl.com

Shiatsu: Therapeutic Art of Japan
www.doubleclicked.com

The Family Village: Cranial Sacral Therapy
www.familyvilage.wisc.edu

The International Center for Reiki Training
www.reiki.org

The Official Pilates Studio™: Pilates™
Method of Body Conditioning
www.pilates-studio.com

The Shiatsu Therapy Association of Australia
www.yogaplace.com.au

**ESSENTIAL OILS & CARRIER OILS**

American Alliance of Aromatherapy
www.205.10.229.2/aaoa/idex/html

Aquaessence Aromatherapy
Body Scrub & Sea Salts
www.purelight.com

Aromatherapy and Essential Oils Information
www.bizbotweekly.com

Aromatherapy Associates
www.cosmed.cnchost.com

Aromatherapy; Oils, Scents, and Information
www.4aromatherapy.com

Aromatherapy—Using Essential Oils
www.sympatico.ca

Atlantic Institute of Aromatherapy
www.atlanticinstitute.com

Bassett Aromatherapy:
World Wide Web Catalog
www.aromaworld.com

Bird's Encyclopedia of Aromatherapy
www.imm.org

Canadian National School of Aromatherapy
www.fragrant.demon.co.uk

Cosmed International
www.cosmed.cnhost.com

Frontier Co-op Aromatherapy
www.frontiercoop.com

Herb Source—Essential Oils and
Aromatherapy Supplies
www.herbsource.com

Herbal Impressions
www.members.aol.com/DBLBFARMS/
index.html

Holistic Health information, The Kevala
Centre, promoting natural healing
www.yoga.kevala.co.uk/products/
aromatherapy_carrier_oils.html

Les Herbes Ltd.—The American Institute for
Aromatherapy and Herbal Studies
www.aromatherapyimst.com

Mountain Breeze
www.mountain.co.uk

Nappies Direct: Aromatherapy Index
www.nappies-direct.co.uk

Poya Essential Oils for Aromatherapy
www.poyanaturals.com

SENSIA Online Shopping:
Aromatherapy, Scented Products and Music
www.sensia.com

The National Association for
Holistic Aromatherapy
www.naha.org

Wicca: Essential Oils & Their Uses
www.odyssy.net/users/erica/wicca/oils.htm

Wildlife Control Technology:
Essential Oils Red Earth:
Use of Essential Oils
www.wctech.com/essoils.htm

Witchware—TripleMoon Witchware:
Pagan, Witch and Wiccan:
Essential Oils
www.witchware.com/essentialoil.htm

**DRIED HERBS**

American Herbs
www.kiva.net
Atlantic Spice Co.
www.atlanticspice.com

European Holistic Aromatherapy
www.hometown.aol.com/aromavitae.
index.html

Fragrant at Demon:
The Guide to Aromatherapy
http://www.fragrant.demon.co.uk

Herbal Impressions
www.members.aol.com/DBLBFARMS/
index.html

Herbs Depot—Herbs, Fragrances,
Spices and Teas
www.herbs-depot.com

Menopause Online
www.menopause-online.com

Michael Tierra's Planetary Herbology,
Acupuncture & Alternative Medicine
www.planetherbs.com

Motherlove Herbals
www.motherlove.com

Swissette Herb Farm
www.meadownet.com/swissette/

Tasmanian Herb Growers Association
www.tassie.net.au/tasherbs

The American Herbal Products Association
www.ahpa.org

The Herb Peddler
www.thefoodstores.com

The Journal of the American Botanical Council
and the Herb Research Foundation
www.healthy.net

**MUD, SEAWEED AND CLAY**
*Just Seacrets*
www.home.earthlink.net


*Sabia ñ Future Nature Pharmacy*
www.abia.com

*The Health Spa*
www.thehealthspa.com


**CANDLES**
*A Lot of Scents—Candles and
Aromatherapy Products*
www.alotofscents.com

*Aroma Ihyme: Spa Treatments,
100% Natural Beeswax Candles*
www.aromathyme.com

*Aromics: The Source for Aromatherapy,
Aromatherapy Candles*
www.aromics.com

*Atlantas Environmental Awarehouse:
The Common Pond, Aromatherapy Candles*
www.thecommonpond.com

*Aurora Ice Aromatherapy Candles*
www.auroraice.com

*Candlexpressó The Candle Shop Online*
www.candlexpress.com


*Earth, Life and More: Aromatherapy, Candles,
Bath Crystals, etc.*
www.earthlifeandmore.com

*Escentials: Aromatherapy Candles*
www.escentials.com

*Ounce of Prevention
Nutrition Shoppe: Candles*
www.ounceofprevention.com

*R.D. Lights Aromatherapy Candles*
www.rdlights.com

*Sensia online shopping: Aromatherapy,
Scented Products and Music*
www.sensia.com

There are over one thousand spas, resorts and health retreats worldwide. So if you positively can't imagine sharing your own bathtub with seaweed, digging your fingers in mud and slathering it all over your body, or if concocting lotions in the kitchen blender does not exactly fit in with your notion of pampering and total relaxation, then have someone else do the work for you, somewhere else, and for a fee.

Your spa selection will primarily depend on three factors: how much you want to spend, how long you want to go for and how far you want to travel, if at all. While all spas basically have pampering in mind, you can further cater to your needs by selecting a spa that pinpoints your ideal based on what programs it offers and what its primary characteristics are.

There are many spa categories, practically one tailored to each and every spa seeker. There are day spas, spas for golfers, for equestrians, for mothers and daughters only, for men only, for chef-wannabes and spas for parents and children. The principal categories, however, can be narrowed down to the following five:

### Ⓕ Fitness/Wellness
These spas specialize in weight and fitness management through exercise and diet. Programs focus primarily on body and mind conditioning.

### Ⓐ Adventure
These spas often encourage mental and physical detoxification through activities such as mountain climbing, trekking, cross-country skiing, snorkeling or yoga.

### Ⓗ New Age/Holistic
These spas promote spiritual growth and emotional balance through meditation, natural remedies and awareness programs.

### Ⓒ Culinary
While most spas offer nutritional plans and guidance in conjunction with body treatments and exercise routines, this category of spas is noted for gourmet meals and innovative cuisine as part of its curative programs.

### Ⓦ Water Treatments
In addition to offering traditional spa remedies, these spas specialize in thalassotherapy and water treatments such as thermal baths and Vichy showers.

# DOMESTIC

**ARIZONA**
Canyon Ranch
8600 E. Rockcliff Rd.
Tucson, AZ 85750
(520) 749-9000
Ⓕ

Marriott's Spa at
Camelback Inn
5402 East Lincoln Drive
Scottsdale, AZ 85253
(602) 905-7021
Ⓕ

Miraval Life in
Balance
5000 E. Via Estancia
Miraval
Catalina, AZ 85739
(520) 825-4000
Ⓕ

The Phoenician
6000 East Camelback Road
Scottsdale, AZ 85251
(602) 423-2405
Ⓕ

**CALIFORNIA**
The Ashram
P.O. Box 8009
Calabassas, CA 91302
(818) 888-0232
Ⓗ

The Chopra Center
for Well Being
7630 Fay Avenue
La Jolla, CA 92037
(888) 424-6772
Ⓗ

The Golden Door
P.O. Box 6305
Escondido, CA 92046-3057
(800) 424-0777
ⒶⒸ

Harbin Hot Springs
P.O. Box 782
Middletown, CA 95461
(707) 982-2477
Ⓗ

La Costa Resort
& Spa
Costa del Mar Road
Carlsbad, CA 92009
(619) 438-9111
Ⓕ (women only)

La Quinta Resort
& Club
P.O. Box 69
La Quinta, CA 92253
(800) 598-3828
ⒻⒶ

The Oaks at Ojai
122 Easr Ojai Avenue
Ojai, CA 93203
(805) 646-5573
ⒶⒽⒸ

Ojai Valley Inn
and Spa
Country Club Road
Ojai, CA 93023
(805) 640-2080
Ⓕ

The Palms
572 North Indian Avenue
Palm Springs, CA 92202
(619) 325-1111
ⒻⒸ

Sonoma Mission Inn
& Spa
18140 Sonoma Hwy. 12
Boyes Hot Springs, CA
95416
(800) 58-9022
ⒻⒸ

Two Bunch Palms
67-425 Two Bunch Palms
Trail
Desert Hot Springs, CA
92240
(800) 472-4334
Ⓦ

White Sulphur Springs
Resort & Spa
3100 White Sulphur
Springs Rd.
St. Helena 94574
(707) 963-4361
Ⓦ

**COLORADO**
The Broadmoor
P.O. Box 1439
Colorado Springs, CO
80901-1439
(719) 634-7711
ⒻⓌ

Global Fitness Adventures
P.O. Box 1390
Aspen, CO 81612
(970) 927-9593
Ⓐ

Gold Lake Mountain Resort
& Spa
3371 Gold Lake Road
Ward, CO 80481
(303) 459-3544
Ⓕ

The Lodge and Spa
at Cordillera
P.O. Box 1110
2205 Cordillera Way
Edwards, CO 81632
(970) 926-2200
Ⓕ

The Peaks
624 Mountain Village Blvd.
P.O. Box 2702
Telluride, CO 81435
(970) 728-6800
Ⓕ

**CONNECTICUT**
Norwich Inn & Spa
607 Thames St., Rt. 32
Norwich, CT 06360
(800) 275-4772
Ⓗ

Saybrook Point Inn
Old Saybrook, CT 06475
(860) 395-2000
Ⓕ

The Spa at Grand Lake
Route 207
Lebanon, CT 06249
(860) 642-4306
Ⓗ

**FLORIDA**
Doral Golf Resort & Spa
8755 N.W. 36th St.
Miami, FL 33178-2401
(305) 593-6030
ⒶⒸ

Eden Roc
4525 Collins Avenue
Miami Beach, FL 33140
(305) 531-0000
Ⓕ

Fisher Island Club
Fisher Island Drive
Fisher Island, FL 33109
(305) 535-6020
Ⓕ

Fit For Life Spa Health
Resort & Spa
1460 S. Ocean Blvd.
Pampano Beach, FL 33062
(954) 941-6688
Ⓕ

Hippocrates Health Institute
1443 Palmdale Court
West Palm Beach, FL 33411
(561) 471-8876
Ⓗ

Palm-Aire
2601 Palm-Aire Drive North
Pompano Beach, FL 33069
(800) 272-5624
Ⓕ

PGA National Resort
& Spa
400 Ave. of the Champions
Palm Beach Gardens, FL
33418
(561) 627-2000
Ⓕ

Safety Harbor Resort
& Spa
105 Bayshore Drive
Safety Harbor, FL 33572
(800) 237-0155
Ⓦ

GEORGIA
The Spa at Chateau Elan
7000 Old Winder Highway
Winder, GA 30680
(770) 307-4433
Ⓗ

Sea Island Spa
at the Cloister
Sea Island GA 31561
(916) 638-3611
Ⓕ

HAWAII
Grand Wailea Resort
Hotel and Spa
3850 Wailea Alanui Drive
Wailea, Maui, HI 96753
(808) 875-1234
Ⓦ

Hilton Waikoloa
425 Waikoloa Beach Drive
Waikoloa, Hawaii 96738
(808) 885-1234
Ⓕ

Hotel Hana-Maui
Wellness Center
P.O. Box 9
Hana, HI 96713
(808) 248-8211

IOWA
The Raj Maharishi
Ayur-Veda Health Center
1734 Jasmine Street
Fairfield, IA 52556-9005
(515) 472-9580
Ⓗ

MAINE
Northern Pines Health
Retreat
559 Route 85
Raymond, ME 04071
(207) 655-7624
ⒶⒽ

MASSACHUSETTS
Canyon Ranch
165 Kemble Street
Lenox, MA
(800) 726-9900
ⒻⒸ

Kripalu Center for
Yoga and Health
30 West Street, Box 793
Lenox, MA 01240
(800) 741-7353
Ⓗ

MINNESOTA
Birdwing Spa
21398 57th Avenue
Litchfield, MN 55255
(320) 693-6064
Ⓐ

NEW MEXICO
Ojo Caliente Mineral
Springs Spa
P.O. Box 68
Ojo Caliente, NM 87549
(505) 583-2233
Ⓦ

Ten Thousand Waves
P.O. Box 10200
Santa Fe, NM 87504
(505) 982-9304
Ⓕ

NEW YORK
Gurney's Inn Resort
and Spa
Route 27, Sunrise Hwy.
Montauk, NY 11954
(516) 668-2345
Ⓦ

New Age Health Spa
P.O. Box 658
Neversink, NY 12765
(800) 682-4348
Ⓗ

NORTH CAROLINA
Structure House Residential
Treatment Center
3017 Pickett Road
Durham, NC
(919) 493-4205
Ⓕ

Westglow Spa
Highway 221 S
Blowing Rock, NC 28605
(704) 2954463
Ⓐ

OHIO
The Kerr House
17777 Beaver Street
P.O. Box 43522
Grand Rapids, OH 43552
(419) 832-1733
Ⓕ (women only)

TEXAS
Cooper Wellness
Program
12230 Preston Road
Dallas, TX 75230
(972) 386-4777
Ⓕ

The Greenhouse
P.O. Box 1144
1171 107th Street
Arlington, TX 76004
(817) 640-4000
(women only)

Lake Austin Spa Resort
1705 South Quinlan Park Rd.
Austin, TX 78732
(800) 847-5737
Ⓐ

UTAH
Green Valley Fitness
Resort & Spa
1871 W. Canyon View Drive
St. George, UT 84770
(801) 628-8060
Ⓕ

Pah Tempe Hot Springs
Resort
825 No 800 East
Hurricane, UT 84737
(435) 635-2879
Ⓦ

VERMONT
Green Mountain at Fox Run
P.O. Box 164
Ludlow, VT 05149
(802) 228-8885
(women only)

New Life Hiking Spa
P.O. Box 395
Killington, VT 05751
(802) 422-4302
Ⓐ

Topnotch at Stowe
Resort & Spa
4000 Mountain Road
Stowe, VT 05672
(800) 451-8686
Ⓕ

WEST VIRGINIA
The Greenbrier
300 West Main Street
White Sulphur Springs, WV
24986
(304) 536-1110
Ⓕ

## INTERNATIONAL
BELGIUM
La Reserve Resort and
Conference Thalassa Beauty
Elisabetlaan 160
Knokke-Heist 8300
(32) 50-61-06-06
Ⓦ

BRAZIL
Caesar Park Pianema
Av. Vieira Souto, 460
22420-000 Rio de Janeiro
021-521-2525
Ⓕ

Maksoud Plaza Sao Paulo
Alameda Campinas, 150
01404-900 Sao Paulo
011-30026
Ⓕ

CANADA
Eco-Med Natural Health Spa
515 Pacific Shores Nature
Resort
Nanoose Bay, BC V9P 9B7
(250) 468-7133
Ⓗ

The Hills Health & Guest
Ranch
Box 26
108 Mile Ranch
British Columbia VOK 2ZO
(250) 791-5225
Ⓐ

The Inn at Manitou
McKeller, Ontario PO6 1CO
(705) 389-2171
Ontario
Ⓕ

Mountain Trek Fitness
Retreat & Health Spa
Box 1352
Ainsworth Hot Springs,
British Columbia
VOG 1AO
(250) 229-5636
Ⓐ

Solace
Box 960
Banff, AB TOL 0CO
(800) 404-1772
Ⓕ

THE CARIBBEAN
Ann Wigmore Institute
P.O. Box 429
Rincon, Puerto Rico 00677
(787) 868-6307
Ⓕ

Ciboney
Box 728
Main Street
Ocho Rios, St. Ann
Jamaica
(800) 777-7800
Ⓕ

Le Sport
P.O. Box 437
Cariblue Beach
St. Lucia
(809) 452-8551
Ⓕ

Jalousie Hilton Resort
& Spa
P.O. Box 251
Soufriere, St. Lucia
(758) 459-7666
Ⓕ

Privilege Resort & Spa
Anse Marcel
St. Martin, French West
Indies 97150
(401) 849-8012
Ⓦ

Sans Souci Lido
P.O. Box 103
Ocho Rios, St. Ann
Jamaica
(800) 467-8737
Ⓕ

FRANCE
Quiberon Thalassotherapy
Institute
B.P. 170
56170 Quiberon
(33-2) 97-50-20-00
ⒸⓌ

Hotel Miramar
13, rue Louison Robert
Biarritz 64200
(33-05) 59-4130-01
Ⓦ

Domaine du Royal Club
Evian
Rive Sud du lac de Geneve
74500 Evian
(33-4) 50-26-8500
Ⓦ

Thalgo-La Baule
Avenue Marie-Louise, B.P. 50
44503 La Baule
(33) 2-4011-9999
Ⓦ

GERMANY
Brenner's Park Hotel
& Spa
An der Lichtentaler Allee
Baden-Baden D-76530
(49) 7221-900-830
ⒶⒸ

Buhlerhohe Schlosshotel
Schwarzwaldhochstrasse 1
Buhl/Baden-Baden 77815
(49) 7226-550
Ⓕ

Clinic Center Bad Sulza
Wunderwaldstrasse 2
Bad Sulza, D-99518
(49) 36461-91826
Ⓗ

GREECE
Royal Mare Thalasso
Limin Hersonnissou
Crete, GR-70014
(30) 623-0400
Ⓦ

HOLLAND
Bilderberg Chateau
Holtmule
Kasteellaan 10
5932 AG Tegelen
077-373-8800
Ⓕ

**HONG KONG**
The Gold Coast Hotel
1 Castle Peak Road
Castle Peak Bay, Kowloon
(852)2452-8888
Ⓕ

**HUNGARY**
Thermal Hotel Margitsziget
Margitsziget, Budapest
H-1138
(36) 1-329-2300
Ⓦ

**IRELAND**
Loughaunrone Health Farm
Rinville West
Oranmore, Co. Galway
(353) 91-790606
Ⓗ

The Merrion Dublin
Upper Merrion Street
Dublin 2
(353) 1-603-0600
Ⓕ

**ITALY**
Albergo Terme San Montano
Lacco Amendo d'Ischia
Napoli 80076
(39) 81-994033
ⒸⓌ

Capri Beauty Farm
Via Capodimonte 2b
Anacapri 80071
(39) 837-3800
Ⓦ

Grotta Guisti Terme Spa
Via Grotta Guisti
Monsummano Terme
171-51015
(39) 572-51008
Ⓦ

Spa'Deus Centro
Benessere
Via Le Piana 35
Chanciano Terme
53042 Tuscany
(39) 578-63232
Ⓕ

Suisse Thermal Village
Via Picota Sentinella
Casamicciola Terme
Island of Ischia (NA) 80074
(39) 081-5980511
Ⓦ

**MEXICO**
Hosteria Las Quintas
Av Diaz Ordaz No 9
C.P. 62440 Cuernavaca, Mor.
(52) 73-183949
Ⓕ

Rio Caliente Spa
P.O. Box 997
Millbrae, CA 94030
(650) 615-9543
ⒶⓌ

Rancho La Puerta
Tecate, Baja
(760) 744-4222
ⒻⒸ

**SCOTLAND**
The Balmoral
1 Princes Street
Edinburgh EH2 2EQ
(44) 01-131-622-8832
Ⓕ

Turnberry Hotel
Golf & Spa
Ayshire, KA26 9LT
(44) 01-655-331-000
Ⓕ

**SPAIN**
Incosol Golf & Spa x
Golf Rio Real
Marbella 29600
(31) 5 282-8500
Ⓕ

Termas de San Roque
Calle San Roque 4
Alhama de Arogon 50230
(34) 976-840014
Ⓦ

Vichy Catalan
Avenida Doctor Furast 32
Caldes de Malavella
Girona 17455
(34) 972-470-000
Ⓦ

Caldas de Besaya
39460 Las Caldas de Besaya
Cantabria
(34) 942-819229
Ⓦ

**SWEDEN**
Selma Lagerlöf Hotel & Spa
2500 P.O. Box 500
28 Sunne S-686
(46) 565-16600
Ⓐ

**SWITZERLAND**
Clinic La Prairie
Clarens-Montreux, 1815
(41) 21-989-3311
Ⓦ

Grand Hotel Park
Wispilestrasse ,
Gstaad CH-3780
(41) 033-748-9800
Ⓕ

Relais and Chateaux Hotel
Giardino
Via Segnale, Ascona,
CH-6612
(41) 91-791-0101
Ⓦ

Victoria-Jungfrau Grand
Hotel and Spa
Hoheweg, Interlaken
CH-3800
(41) 033-828-2828
Ⓕ

Grand Hotel des Bains
Avenue des Bains 22
Yverdon-Les Bains, CH-1400
(41) 24-425-7021,
Ⓦ

**THAILAND**
Banyan Tree Phuket Spa
Resort
Phuket
(66) 62-361771210
Ⓦ

Chiva-Som
73 Petchkasem Road
Hua Hin 77110
(66) 32-536536
Ⓕ

**UNITED KINGDOM**
Grayshott Hall
Health Fitness Retreat
Headley Road, Grayshott
Surrey
(44) 01-428-604-331
Ⓕ

Henlow Grange
Health Farm
Henlow, Bedforshire
SG16 6DB
(44) 01-462-811-111
Ⓕ

Springs Hydro
Packington,
Nr. Ashby de la Zouch, Leicestershire
LE6 51TG
(44) 1-530-273-873
Ⓦ

VENEZUELA
La Samanna Hotel &
Thalassotherapy Center
Av. Bolivar con Av. Gomez
Porlamar Urb. Costa Azul
(58) 95-62-26-62
Ⓦ

## SHOPPING GUIDE

**Aveda**
509 Madison Avenue
New York, NY 10022
(212) 832-2416
or for retail store nearest you call
(800) AVEDA-24

**Aromatherapy Products**
P.O. Box 2354
Fair Oaks, CA 95628
(916) 965-7546

**Aroma Vera**
5901 Rodeo Rd.
Los Angeles, Ca 90016-4312
essential oils, aromatherapy supplies,
floral waters

**American Botanical Council**
P.O. Box 2016600
Austin, Texas 78720-1660
(800) 373-7105
for quarterly magazine,
herbal education catalog

**Atlantic Spice Company**
P.O. Box 205
N. Truri, MA 02652
(800) 316-7965
for herbs and spices

**The Body Shop International Ltd**
Hawthorn Road
Little Hampton,
West Susex BN17 7LR,
England
44-0903-726-250

**The Body Shop by Mail**
45 Horsehill Road
Cedar Knolls, NJ 07927
(800) 541-2535

**The Essential Oil Company**
P.O. Box 206
Lake Oswego, OR 97034
(800) 729-5912
essential oils

**Herbal Home Store**
775 Santa Cruz Avenue
Menlo Park, CA 94025
(650)325-7848
(888)325-7848

**Green Terrestrial Herbal Products**
(203) 862-8690

**Gingham 'n' Spice, Ltd.**
P.O. Box 88psc
Gardenwille, PA 18926
(215) 348-3585
carrier oils, essential oils,
bottles, beeswax pearls,
aloe gel, rosewater,
glycerin, glass rods, droppers

**Kiehl's Pharmacy**
109 Third Avenue
New York, NY 10003
(212) 475-3400

**Lavender Lane**
7337 # 1Roseville Road
Sacramento CA 95842
(888) 593-4400

**Mountain Rose Herbs**
P.O. Box 2000
Redway, California 95560
herbs, oils bottles, clays, floral waters

**Margaret's Magicals**
P.O. Box 846
Planertarian Station
(212) 496-6726

**Naomi's Herbs**
11 Housatonic Street
Lenox, MA 02140
(888) 462-6647
herbs, aromatherapy products, capsules,
bottles

Begoun, Paula. *The Beauty Bible*. Seattle, Washington: Beginning Press, 1997.

Benson, Elaine. *Unmentionables: A Brief History of Underwear*. New York: Simon & Schuster, 1996.

Breedlove, Greta. *The Herbal Home Spa*. Vermont: Storey Books, 1998.

Capellini, Steve. *The Royal Treatment*. New York: Dell, 1997.

Carpenter, Deb. *Nature's Beauty Kit*. Colorado: Fulcrum Publishing, 1995.

Carter, Alison. *Underwear: The Fashion History*. New York: Drama Book Publishers, 1992.

Chase, Deborah. *The New Medically Based No-Nonsense Beauty Book*. New York: Henry Holt and Company, 1989.

Cooper, Wendy. *Hair: Sex, Society, Symbolism*. New York: Stein and Day Publishers, 1971.

Cox, Janice. *Natural Beauty at Home*. New York: An Owl Book/Henry Holt and Company, 1995.

Cox, Janice. *Natural Beauty for All Seasons*. New York: An Owl Book/Henry Holt and Company, 1996.

Doner, Kalia. *The Wellness Center's Spa at Home*. New York: Berkley Books, 1997.

Fontanel, Beatrice. *Support & Seduction: A History of Corsets and Bras*. New York: Harry N. Abrams, Inc., 1997.

Gizowska, Eva. *Reader's Digest Bathing for Health, Beauty, & Relaxation*. Pleasantville: The Reader's Digest Association, Inc., 1998.

Groom, Nigel. *The Perfume Handbook*. New York: Chapman & Hall, 1992.

Harding, Anne. *Select Spas*. New York: McGraw Hill, 1989.

Harris, Jessica B. *The World Beauty Book*. San Francisco: Harper San Francisco, 1995.

Joseph, Jeffrey. *Spa-Finders Guide to Spa Vacations At Home and Abroad.* New York: John Wiley & Sons, Inc., 1990.

Kelar, Casey. *The Natural Beauty & Bath Book.* Ashville, NC: Lark Books, 1997.

Keller, Erich. *The Complete Home Guide to Aromatherapy.* Tiburon, CA: H J Kramer, Inc., 1991.

Lakoff, Robin and Scherr, Raquel L. *Face Value—The Politics of Beauty.* Boston: Routledge & Kegan Paul, 1984.

Lawless, Julia. *The Illustrated Encyclopedia of Essential Oil.* Rockport, MA: Element Books, 1995.

Leigh, Michelle Dominique. *The Japanese Way of Beauty.* Secaucus, NJ: Birch Lane Press, 1992.

Maloney, Kathleen. *The Canyon Ranch Health and Fitness Program.* New York: Simon and Schuster, 1989.

Riggs, Maribeth. *The Scented Woman.* New York: Viking Studio Books, 1992.

Ryman, Daniele. *Aromathcrapy—-The Complete Guide to Plant and Flower Essences for Health and Beauty.* New York: Bantam Books, 1993.

Sarnoff, Pam Martin. The Ultimate Spa Book. New York: Warner Books, 1989.

Spiers, Katie. *Recipes for Natural Beauty.* New York: Facts on File, Inc., 1998.

Webb, David. *Making Potpourri Colognes and Soaps.* Tab Books, 1988.

Weinberg, Norma Pasekoff. *Natural Hand Care Book.* Pownal VT: Storey Books, 1998.

## CONVERSION TABLE

| This much | Equals approximately this much* |
|---|---|
| 1 milliliter | 25 drops or<br>1/5 teaspoon or<br>1/15 tablespoon or<br>1/30 fluid ounce |
| 1 teaspoon | 100 drops or<br>5 milliliters or<br>1/3 tablespoon or<br>1/6 fluid ounce |
| 1 tablespoon | 300 drops or<br>15 milliliters or<br>3 teaspoons or<br>1/2 fluid ounce or<br>1/16 cup |
| 1 fluid ounce | 30 milliliters or<br>6 teaspoons or<br>2 tablespoons or<br>1/8 cup or<br>1/16 pint or<br>1/32 quart or<br>1/32 liter |
| 1 pint | 474 milliliters<br>16 fluid ounces or<br>2 cups or<br>1/2 quart or<br>1/2 liter or<br>1/8 gallon |
| 1 quart | 32 fluid ounces or<br>4 cups or<br>2 pints or<br>1 liter or<br>1/4 gallon |
| 1 gallon | 128 fluid ounces or<br>16 cups or<br>8 pints or<br>4 liters |

**\*Note:** These measurements are rounded off; while they are not precise, they are sufficiently accurate for the purposes of the measurements required for the recipes in this book.

## INDEX

### A

Acne, *see Blemishes.*
Almond oil, 40, 89, 102, 105, 111, 121
Almonds, 15, 125, 153
Aloe vera gel, 43, 81, 87, 103, 127
Apricot kernel oil, 17
Apricot Mousse, 157
Apricot oil, 51, 116, 127
Apricots, 127, 157;
Arugula, 155
Avocado oil, 117
Avocados, 15, 83, 121, 134

### B

Backache relief, 35, 61
Bananas, 91, 149, 150
Basil, 31
Baths,
  essential oil 32-45;
  herbal, 27-31
Beans,
  string, 154;
  white, 154
Beeswax, 49, 51, 53, 103, 127
Beets, 154
Bergamot oil, 39, 77
Blemishes, 77, 99
Blueberries, frozen, 150
Body wraps, 19-25
Buttermilk, 71

### C

Cantaloupe, 149
Carrier oils, 166-167;
  *see also Almond oil;*
  *Apricot oil; Apricot kernel*
  *oil; Avocado oil; Castor oil;*
  *Coconut oil; Grapeseed oil;*
  *Hazelnut oil; Jojoba oil;*
  *Olive oil; Sesame oil; Soy*
  *oil; Sunflower oil; Wheat*
  *germ oil.*
Carrots, 83, 153
Castile soap, 139, 140
Castor oil, 113
Cedarwood oil, 61
Celery, 152
Chamomile, 20-21;
  Roman, 139
Cherries, 156
Chickpea flour, 91
Chickweed, 119

Clary sage oil, 57
Clay,
  green, 122;
  white, 133
Cloves, 140
Cocoa butter, 49, 125
Coconut,
  oil, 49, 101, 133;
  shredded, 153
Cold cream, 102
Comfrey, 30
Cornmeal, 117
Cornmeal & Pumice
  Foot Scrub, 117
Cottage cheese, 157
Cranberries, 99;
  juice, 151
Cream, fresh, 39;
  heavy, 83
Crisco, 69
Cucumbers, 87, 97, 102, 118, 152, 153
Cypress oil, 39

### D

Deodorant, foot, 122
Desserts, 156-157
Detoxification, 25, 27, 81
Dill, 155
Drinks, 149 151, 156

### E

Eucalyptus, dried, 31;
  oil, 35
Eggs, 101;
  whites, 78
Endives, Belgian, 155
Essential oils, 41, 160-165; *see also*
  *Bergamot oil; Cedarwood*
  *oil; Clary sage oil; Cypress*
  *oil; Eucalyptus oil;*
  *Galbanum oil; Geranium*
  *oil; Juniper oil; Lavender*
  *oil; Lemon oil; Neroli oil;*
  *Orange oil; Patchouli oil;*
  *Peppermint oil; Rose oil;*
  *Rosemary oil; Sandalwood*
  *oil; Tea tree oil;*
  *Thyme oil; Vetiver oil;*
  *Ylang-ylang oil.*
Exfoliation, 10, 17, 71, 116, 117, 125;
  *see also Scrubs.*

### F

Facial cleansers, 64-71;
  masks & peels, 78-93;
  steams, 72-77;
  toners, 94-99
Fennel seeds, 67

### G

Galbanum oil, 51
Gazpacho, 152
Gelatin, 87, 93, 157
Geranium oil, 37, 41
Ginger, 25, 29
Grape juice, white, 150
Grapeseed oil, 57, 119, 136

### H

Hair, 130-131;
  brittle, 136; 142;
  dry, 134, 141;
  normal, 135;
  pre-shampoo treat-
  ments, 133-137;
  rinses, 143-145;
  shampoos &
  conditioners, 139-141
Haricots verts, 154
Hazelnut oil, 105
Hazelnuts, 125
Henna powder,
  neutral, 142
Honey, 39, 71, 83, 89, 111, 113, 125, 137, 149, 150, 151, 156, 157
Honeydew melons, 150
Hydrogen peroxide, 111

### J

Joint pain relief, 61
Jojoba oil, 59, 61, 111, 133, 139, 140
Juniper oil, 37

### K

Kelp, *see Seaweed.*
Kelp powder, 81
Kiwis, 151

### L

Lanolin, 102, 103
Lavender, 20-21, 120;
  oil, 33, 45, 57, 77, 119, 120, 135
Lecithin, liquid, 101

Lemon oil, 53
Lemons, 151;
  juice, 67, 97, 101, 116;
  peel, 29, 45
Lemongrass, 25, 30
Lentils, 153
Limes, 150, 151;
  juice, 152, 157
Linden, 28

**M**
Maple syrup, 137
Massage oils, 56-61
Mayonnaise, 134
Milk, dried, 45
Moisturizers,
  body, 47-53;
  facial, 101-105;
  hand, 127
Molasses, 137
Mood enhancers, 170
Mustard, Dijon, 155

**N**
Nails & cuticles,
  108-109;
  treatments, 111-112;
  whitener, 111
Neroli oil, 33, 39

**O**
Oatmeal, 15, 28, 29, 31,
  71, 121, 125
Oils, *see Carrier oils;*
  *Essential oils;* Names of
  specific oils.
Olive oil, 41, 49,
  85, 103, 111, 112,
  113, 121, 135,
  153, 154, 155
Onions, red, 152,
  153, 155
Orange oil, 39, 105
Oranges, 151, 156;
  juice, 149;
  peel, 30, 45

**P**
Papaya, 93, 157
Parsley, 29, 153, 154
Patchouli oil, 139, 140
Pears, Bosc, 155
Peppers,
  red bell, 152;
  yellow, 154
Peppermint, 31;

oil, 75, 124
Pineapples, 85
Pineapple & Olive Oil
  Mask, 85
Potatoes, Red Bliss, 155
**R**
Raisins, 153
Rejuvenation, 25
Relaxation, 20, 27,
  28, 37
Remineralization, 23, 81
Revitalization, 13, 29
Rose, oil, 51, 89;
  petals, 73, 93;
  water, 51, 73, 94, 103,
Rosemary, 31, 140, 143;
  oil, 13, 33, 75

**S**
Salads, 153-155
Sandalwood oil, 40
Sage, 31, 120, 136, 140
Savory, dried, 99
Scallions, 152
Scrubs, body, 10-17;
  hand & foot, 113,
  116-117, 121, 125;
  *see also Exfoliation.*
Sea salt, 13, 17
Seaweed, 23;
  hijiki, powdered, 145;
  powdered, 43
Sesame oil, 49, 101,
  139, 140
Shallots, 154
Skin,
  aging, 89, 95;
  dry, 40, 47, 49, 101;
  normal, 41, 47, 51,
  67, 73, 97;
  oily, 39, 47, 51, 69,
  77, 97
Sour cream, 142
Soy oil, 124
Sugar, 116
Sunflower oil, 101

**T**
Tea, chamomile, 87;
  green, 87, 95;
  orange pekoe, 156;
  peppermint, 97
Tea tree oil, 141
Thyme, 61; oil, 35, 61
Tools, 169-170
Tomatoes, 67, 152, 153,

cherry, 154

**V**
Valerian, 20-21
Vanilla extract, 156
Vetiver oil, 59
Vinaigrette,
  balsamic, 153;
  dill, 155
Vinegar,
  apple cider, 69, 143;
  balsamic, 152, 153,
  154, 155;
  red wine, 153;
  white, 109
Vitamin A, 15, 17, 65
Vitamin B, 17
Vitamin D 15
Vitamin E, 15, 65;
  capsules, 111, 127;
  oil, 111, 112

**W**
Walnuts, 113
Washes, hand, 118-120,
  122-125
Watermelons,
  yellow, 152
Wheat germ oil, 53, 136
Witch hazel, 53, 97, 118
Wraps, body, 19-25
Wrinkles, 105

**Y**
Yarrow, 28
Ylang-ylang oil, 39, 139
Yogurt, frozen,
  vanilla, 149;
  plain, 11, 142,
  156, 157